D0436677

WITHDRAWN
FEB 01 2023

MICHAEL WHITE AND JOHN GRIBBIN

STEPHEN HAWKING

A LIFE IN SCIENCE

PENGUIN BOOKS

PENGUIN BOOKS

Published by the Penguin Group
Penguin Books Ltd, 27 Wrights Lane, London W8 5TZ, England
Penguin Books USA Inc., 375 Hudson Street, New York, New York 10014, USA
Penguin Books Australia Ltd, Ringwood, Victoria, Australia
Penguin Books Canada Ltd, 10 Alcorn Avenue, Toronto, Ontario, Canada M4V 3B2
Penguin Books (NZ) Ltd, 182–190 Wairau Road, Auckland 10, New Zealand

Penguin Books Ltd, Registered Offices: Harmondsworth, Middlesex, England

First published by Viking 1992
Published in Penguin Books 1992
7 9 10 8 6

Printed in England by Clays Ltd, St Ives plc

Contents

Preface

When Stephen Hawking was involved in a minor road accident in Cambridge city centre early in 1991, within twelve hours American TV networks were on the phone to his publisher, Bantam, for a low-down on the story. The fact that he suffered only minor injuries and was back at his desk within days was irrelevant. But then, anything about Stephen Hawking is newsworthy. This would never have happened to any other scientist in the world. Apart from the fact that physicists are seen as somehow different from other human beings, existing outside the normal patterns of human life, there is no other scientist alive as famous as Stephen Hawking.

But Stephen Hawking is no ordinary scientist. His book *A Brief History of Time* has notched up worldwide sales in the millions – publishing statistics usually associated with the likes of Jeffrey Archer and Stephen King. What is even more astonishing is that Hawking's book deals with a subject so far removed from normal bedtime reading that the prospect of tackling such a text would send the average person into a paroxysm of inadequacy. Yet, as the world knows, Professor Hawking's book is a massive hit and has made his name around the world. Somehow, he has managed to circumvent prejudice and to communicate his esoteric theories directly to the lay-reader.

However, Stephen Hawking's story does not begin or end

with *A Brief History of Time*. First and foremost, he is a very fine scientist. Indeed, he was already established at the cutting edge of theoretical physics long before the general public was even aware of his existence. His career as a scientist began over twenty-five years ago when he embarked on cosmological research at Cambridge University.

During those twenty-five years, he has perhaps done more than anyone to push back the boundaries of our understanding of the Universe. His theoretical work on black holes and his progress in advancing our understanding of the origin and nature of the Universe have been ground-breaking and often revolutionary.

As his career has soared, he has led a domestic life as alien to most people as his work is esoteric. At the age of twenty-one Hawking discovered that he had the wasting disease ALS, also called motor neuron disease, and he has spent much of his life confined to a wheelchair. However, he simply has not allowed his illness to hinder his scientific development. In fact, many would argue that his liberation from the routine chores of life has enabled him to make greater progress than if he were able-bodied.

Stephen Hawking does not like to dwell too much on his disabilities, and even less on his personal life. He would rather people thought of him as a scientist first, popular science writer second, and, in all the ways that matter, a normal human being with the same desires, drives, dreams and ambitions as the next person. In this book we have tried our best to respect his wishes and have endeavoured to paint a picture of a man with talents in abundance, but none the less a man like any other.

In attempting to describe Professor Hawking's work as well as the life of the man behind the science, we hope to enable the reader to see both from different perspectives. Although there are inevitable overlaps in the story, we hope this will help to place the science within the human context – indeed, to show that, for Stephen Hawking, science and life are inextricably linked.

Michael White, Oxford
John Gribbin, Lewes
July 1991

Acknowledgements

We would like to thank a number of people who, for one reason or another, helped to make this book happen: Mark Barty-King, Dr Robert Berman, Maureen Berman, Roberta Bernstein, staff at the Cambridge County Library, Professor Brandon Carter, Marcus Chown, Michael Church, Virgil Clarke, Sami Cohen, Dr Kevin Davies, Professor Paul Davies, Sue Davies, Fischer Dilke, Norman Dix, Dr Fay Dowker, Professor George Efstathiou, Professor George Ellis, Peter Guzzardi, Professor Edward Harrison, Professor Stephen Hawking, David Hickman, Chris Holifield, Professor Maurice Jacob, Dr David Lindley, Shirley MacLaine, Dr John McClenahan, Ravi Mirchandani, Dr Simon Mitton, Dr Joseph Needham, Professor Don Page, Murray Pollinger, Colonel Geoffrey Pryke OBE, Professor Abdus Salam, Professor David Schramm, Professor Dennis Sciama, Lydia Sciama, Professor Paul Steinhardt, Rodney Tibbs, Professor Michael Turner, Dr Tanmay Vachaspati, Professor Alex Vilenkin, Lisa Whitaker, Nigel Wood-Smith.

1

The Day Galileo Died

In an upmarket restaurant near Cambridge city centre, twelve young men and women sit around a large, linen-covered table set with plates and dishes, glasses and cutlery. To one side is a man in a wheelchair. He is older than the others. He looks terribly frail, almost withered away to nothing, slumped motionless and seemingly lifeless against the black cloth cushion of his wheelchair. His hands, thin and pale, the fingers slender, lie in his lap. Set into the centre of his sinewy throat, just below the collar of his open-necked shirt, is a plastic breathing device about two inches in diameter. But despite his disabilities, his face is alive and boyish, neatly brushed brown hair falling across his brow, only the lines beneath his eyes belying the fact that he is a contemporary of Keith Richards and Donald Trump. His head lolls forward, but from behind steel-rimmed spectacles his clear blue eyes are alert, raised slightly to survey the other faces around him. Beside him sits a nurse, her chair angled towards his as she positions a spoon to his lips and feeds him. Occasionally she wipes his mouth.

There is an air of excitement in the restaurant. Around this man the young people laugh and joke, and occasionally address him or make a flippant remark in his direction. A moment later the babble of human voices is cut through by a rasping sound, a metallic voice, like something from the set of *Star Wars* – the man in the wheelchair makes a response which brings peals of laughter from the whole table. His eyes light up, and what has been described by some as 'the greatest smile in the world' envelops his whole face. Suddenly you know that this man is very much alive.

As the diners begin their main course there is a commotion at the restaurant's entrance. A few moments later, the head waiter walks towards the table escorting a smiling redhead in a fake-fur coat. Everyone at the table turns her way as she approaches and there is an air of hushed expectation as she smiles across at them and says, 'Hello,' to the gathering. She appears far younger than her years and looks terribly glamorous, a fact exaggerated by the general scruffiness of the young people at the table. Only the older man in the wheelchair is neatly dressed, in a plain jacket and neatly pressed shirt, his immaculately smart nurse beside him.

'I'm so sorry I'm late,' she says to the party. 'My car was wheel-clamped in London.' Then she adds, laughing, 'There must be some cosmic significance in that!'

Faces look towards her and smile, and the man in the wheelchair beams. She walks around the table towards him, as his nurse stands at his side. The woman stops two steps in front of the wheelchair, crouches a little and says, 'Professor Hawking, I'm delighted to meet you. I'm Shirley MacLaine.'

He smiles up at her and the metallic voice simply says, 'Hello.'

For the rest of the meal Shirley MacLaine sits next to her host, plying him with question after question in an attempt to discover his views on subjects which concern her deeply. She is interested in metaphysics and spiritual matters. Having spoken to holy men and teachers around the world, she has formulated her own personal theories concerning the meaning of existence. She has strong beliefs about the meaning of life and the reason for our being here, the creation of the Universe and the existence of God. But they are only beliefs. The man beside her is perhaps the greatest physicist of our time, the subjects of his scientific theories the origin of the Universe, the laws which govern its existence and the eventual fate of all that has been created – including you, me and Ms Shirley MacLaine. His fame has spread far and wide, his name known by millions around the world. She asks the professor if he believes that there is a God who created the Universe and guides His creation. He smiles momentarily, and the machine voice says, 'No.'

The professor is neither rude nor condescending; brevity is simply his way. Each word he says has to be painstakingly spelt out on a computer attached to his wheelchair and operated by tiny movements of two of the fingers of one hand, almost the last vestige of bodily freedom he has. His guest accepts his words and nods. What he is saying is not what she wants to hear, and she does not agree – but she can only listen and take note, for, if nothing else, his views have to be respected.

Later, when the meal is over, the party leaves the restaurant and returns to the Department of Applied Mathematics and Theoretical Physics at the University, and the two celebrities are left alone with the ever-present nurse in Professor Hawking's office. For the next two hours, until tea is served in the common-room, the Hollywood actress asks the Cambridge professor question after question.

By the time of their encounter in December 1988, Shirley MacLaine had met many people, the great and the infamous. Several times nominated for an Oscar, and winner of one for her role in *Terms of Endearment*, she was probably a more famous name than her host that day. Doubtless, though, her meeting with Stephen Hawking will remain one of the most memorable of her life. For this man, weighing no more than ninety pounds and completely paralysed, speechless and unable to lift his head should it fall forward, has been proclaimed 'Einstein's heir', 'the greatest genius of the late twentieth century', 'the finest mind alive' and even, by one journalist, 'Master of the Universe'. He has made fundamental breakthroughs in cosmology and, perhaps more than anyone else alive, he has pushed forward our understanding of the Universe we live in. If that were not enough, he has won dozens of scientific prizes. He has been made a CBE – Commander of the British Empire – and then Companion of Honour by Queen Elizabeth II, and has written a popular science book, *A Brief History of Time*, which has not left the British best-sellers list since its publication in 1988, and has to date sold over ten million copies worldwide.

How did all this happen? How has a man with a progressive wasting disease fought off the ravages of his disability to overcome every obstacle in his path and win through? How has he managed to achieve far more than the vast majority of able-bodied people would ever have dreamt of accomplishing?

To casual visitors, the city of Oxford in January 1942 would have appeared little changed since the outbreak of the Second World War, two and a half years earlier. Only upon closer inspection would they perhaps have noticed the gun emplacements dotted around the city, the fresh camouflage paint in subdued khaki and grey, the high towers protruding from the car plants at Cowley, east of the dreaming spires, and the military trucks and personnel carriers periodically trundling over Magdalen Bridge and along The High, where frost lingered on the stone gargoyles.

Out in the wider world, the war was reaching a crucial stage. A month earlier, on 7 December, the Japanese had attacked Pearl Harbor and the USA had joined the war. To the east the Soviet army was fighting back Hitler's troops in the Crimea, bringing about the first moves that would eventually precipitate the total defeat of both Germany and Japan.

In Britain every radio was tuned to J.B. Priestley presenting *Post-Scripts to the News*; there were Dr Joad and Julian Huxley arguing over trivia and homely science on the 'Brains Trust'; and the 'Forces' sweetheart', Vera Lynn, was wowing the troops at home and abroad with 'We'll Meet Again'. Winston Churchill had just returned from his Christmas visit to America where he had addressed both Houses of Congress, rousing them with quotes from Lincoln and Washington and waving the V-sign. Television was little more than a laboratory curiosity.

It is perhaps one of those oddities of serendipity that 8 January 1942 was both the three-hundredth anniversary of the death of one of history's greatest intellectual figures, the Italian scientist Galileo Galilei, and the day Stephen William Hawking was born into a world torn apart by war and global strife. But, as Hawking himself points out, around two hundred thousand other babies were born that day, so maybe it is after all not such an amazing coincidence.

Stephen's mother, Isobel, had arrived in Oxford only a short time before the baby was due. She lived with her husband Frank in Highgate, a northern suburb of London, but they had decided that she should move to Oxford to give birth. The reason was simple. Highgate, along with the rest of London and much of southern England, was being pounded by the German *Luftwaffe* night after night. However, the warring governments, in a rare display of equanimity, had agreed that if Germany refrained from bombing Oxford or Cambridge, the RAF would guarantee peaceful skies over Heidelberg and Göttingen. In fact, it has been said that Hitler had earmarked Oxford as the prospective capital of world government when his imagined global conquest had been accomplished, and that he wanted to preserve its architectural splendour.

Both Frank and Isobel Hawking had been to Oxford before – as students. They both came from middle-class families. Frank Hawking's grandfather had been quite a successful

Yorkshire farmer, but had seen his prosperity disappear in the great agricultural depression which immediately followed the First World War. Isobel, the second-eldest of seven, was the daughter of a doctor in Glasgow. Neither family could afford university fees without making sacrifices, and in an age where far fewer women went on to higher education than we are now accustomed to, it demonstrated considerable liberalism on Isobel's parents' part that a university education was considered at all.

Their paths never crossed at Oxford, as Frank Hawking went up before his future wife. He studied medicine and became a specialist in tropical diseases. The outbreak of hostilities in 1939 found him in East Africa studying endemic medical problems. When he heard about the war he decided to set off back to Europe, travelling overland across the African continent and then by ship to England, with the intention of volunteering for military service. However, upon arriving home he was informed that his skills would be far more usefully employed in medical research.

After leaving Oxford, Isobel had stumbled into a succession of loathed jobs, including a spell as an inspector of taxes. Leaving after only a few months, she decided to take a job for which she was ridiculously overqualified – as a secretary at a medical research institute. It was there that the vivacious and friendly Isobel, mildly amused at the position she had found herself in but with sights set on a more meaningful future, first met the tall, shy young researcher fresh back from exciting adventures in exotic climes.

When he was two weeks old, Isobel Hawking took Stephen back to London and the raids. They almost lost their lives when he was two, when a neighbour's house was hit by a V2 rocket. Although their home was damaged, the Hawkings were out at the time.

After the war, Frank Hawking was appointed Head of the Division of Parasitology at the National Institute of Medical

Research. The family stayed on in the house in Highgate until 1950, when they moved twenty miles north to a large rambling house at 14 Hillside Road in the city of St Albans in Hertfordshire.

St Albans is a small city dominated by its cathedral, which can trace its foundation back to the year AD 303, when St Alban was martyred and a church built on the site. However, long before that the strategically useful position of the area had been realized by the Romans. There they built the city of Verulamium, and the first Christian church was probably constructed from the Roman ruins left behind when the empire began to crumble and the soldiers returned home. In the 1950s, St Albans was an archetypal, prosperous, middle-class English town. In the words of one of Hawking's school-friends, 'It was a terribly smug place, upwardly mobile, but so awfully suffocating.'

Hawking was eight when the family arrived there. Frank Hawking had a strong desire to send Stephen to a private school. He had always believed that a private-school education was an essential ingredient for a successful career. There was plenty of evidence to support this view: in the 1950s, the vast majority of Members of Parliament had enjoyed a privileged education, and most senior figures in institutions such as the BBC, the armed forces and the country's universities had been to private schools. Dr Hawking himself had attended a minor private school, and he felt that even with this semi-élite background he had still experienced the prejudice of the establishment. He was convinced that, coupled with his own parents' lack of money, this had held him back from achieving greater things in his own career, and that others with less ability but more refined social mores had been promoted ahead of him. He did not want this to happen to his eldest son. Stephen, he decided, would be sent to Westminster, one of the best schools in the country.

When he was ten, the boy was entered for the Westminster

School scholarship examination. Although his father was doing well in medical research, a scientist's salary could never hope to cover the school fees at Westminster – such things were reserved for the likes of admirals, politicians and captains of industry. Stephen had to be accepted into the school on his own academic merit; he would then have his fees paid, at least in part, by the scholarship. The day of the examination arrived and Stephen fell ill. He never sat the entrance paper, and consequently never obtained a place at one of England's best schools.

Disappointed, Dr Hawking enrolled his son at the local private school, St Albans School, a well-known and academically excellent abbey school which had close ties with the cathedral extending back, according to some accounts, to the year AD 948. Situated in the heart of the city and close to the cathedral, St Albans School had 600 boys when Stephen arrived there in September 1952. Each year was streamed as A, B or C according to academic ability. Each boy spent five years in senior school, progressing from the first form to the fifth, at the end of which period he would sit Ordinary (O) Level exams in a broad spectrum of subjects, the brighter boys taking eight or nine examinations. Those who were successful at O Level would usually stay on to sit Advanced (A) Levels in preparation for university two years later.

In 1952 there were on average three applicants for every place at St Albans School and, as with Westminster, each prospective candidate had to take an entrance examination. Stephen was well prepared. He passed easily and, along with exactly ninety other boys, was accepted into the school on 23 September 1952. The fees were fifty-one guineas (£53.55) a term.

The image of Stephen at this time is that of the schoolboy swot in his grey school uniform and cap as caricatured in the 'Billy Bunter' stories and *Tom Brown's Schooldays*. He was eccentric and awkward, skinny and puny. His school uniform always looked a mess and, according to friends, he jabbered rather than talked clearly, having inherited a slight lisp from his father. His friends dubbed his speech 'Hawkingese'. All this had nothing to do with any early signs of illness; he was just that sort of kid – a figure of classroom fun, teased and occasionally bullied, secretly respected by some, avoided by most. It appears that at school his talents were open to some debate: when he was twelve, one of his friends bet another a bag of sweets that Stephen would never come to anything. As Hawking himself now says modestly, 'I don't know if this bet was ever settled, and, if so, which way it was decided.'[1]

By the third year Stephen had come to be regarded by his teachers as a bright student, but only a little above average in the top class in his year. He was part of a small group who hung around together and shared the same intense interest in their work and pursuits. There was the tall, handsome figure of Basil King, who seems to have been the cleverest of the group, reading Guy de Maupassant at the age of ten and enjoying opera while still in short trousers. Then there was John McClenahan, short, with dark-brown hair and a round face, who was perhaps Stephen's best friend at the time. Fair-haired Bill Cleghorn was another of the group, completed by the energetic and artistic Roger Ferneyhaugh, and a newcomer in the third form, Michael Church. Together they formed the nucleus of the brightest of the bright students in class 3A.

The little group were definitely the smart kids of their year. They all listened to the BBC's Third Programme on the radio, now known as Radio 3, which played only classical music. Instead of listening under the sheets to early rock'n'roll or the latest cool jazz from the States, Mozart, Mahler and Beethoven would trickle from their radios to accompany last-minute physics revision for a test the next day or the geography homework due in by the morning. They read

Kingsley Amis and Aldous Huxley, John Wyndham, C.S. Lewis and William Golding – the 'smart' books. Pop music was on the other side of the 'great divide', *infra dig*, slightly vulgar. They all went to concerts at the Albert Hall. A few of them played instruments, but Stephen was not very dextrous with his hands and never mastered a musical instrument. The interest was there but he could never progress beyond the rudiments, a source of great regret throughout his life. Their shared hero was Bertrand Russell, at once intellectual giant and liberal activist.

St Albans School proudly boasted a very high intellectual standard, a fact recognized and appreciated by the Hawkings very soon after Stephen started there. Before long, any nagging regrets that he had been unable to enter Westminster were forgotten. St Albans School was the perfect environment for cultivating natural talent.

Much remembered and highly thought of was a master fresh out of university called Finlay who, way ahead of his time, taped radio programmes and used them as launch points for discussion classes with 3A. The subject-matter ranged from nuclear disarmament to birth control, and everything in between. By all accounts, he had a profound effect on the intellectual development of the thirteen-year-olds in his charge, and his lessons are still fondly remembered by the journalists, writers, doctors and scientists they have become today.

They were forever bogged down with masses of homework, usually three hours each night, and plenty more at weekends, after Saturday morning lessons and compulsory games on Saturday afternoons. Despite the pressures, they still managed to find a little time to see each other out of school. Theirs was pretty much a monastic lifestyle. English schoolboys attending the private schools of the 1950s had little time for girls in their busy programme, and parties were single-sex affairs until the age of fifteen or sixteen. It was only then that they would have the inclination and parental permission to hold sherry parties at their houses, and practise the dance steps they had learnt after school games on Saturdays at a dance studio in St Albans city centre.

Until they had graduated to such pleasures, the boys often went on long bicycle rides in the Hertfordshire countryside around St Albans, sometimes going as far afield as Whipsnade, some fifteen miles away. Another favourite hobby was inventing and playing board games. The key characters in all this were Stephen and Roger Ferneyhaugh. Hawking, the embryonic scientist and logician already emerging, would devise the rules and laws of the games, while Ferneyhaugh designed the boards and pieces. The group would gather at parents' houses during school holidays and at weekends, and set up the latest game on the bedroom floor or with glasses of orange squash on the sitting-room carpet.

First there was the War Game, based on the Second World War. Then came the Feudal Game, devised around the social, military and political intricacies of medieval England, with the whole infrastructure meticulously developed. However, it soon became apparent that there was a major flaw in their games – Stephen's rules were of such labyrinthine complexity that the enactment and consequences of a single move turned out to be so convoluted that sometimes a whole afternoon would be spent sorting them out. Often the games moved to 14 Hillside Road, and the boys would traipse up the stairs to Stephen's cluttered bedroom near the top of the house.

By all accounts the Hawkings' home was an eccentric place, clean but cluttered with books, paintings, old furniture and strange objects gathered from various parts of the world. Neither Isobel nor Frank Hawking seemed to care too much about the state of the house. Carpets and furniture remained in use until they began to fall apart; wallpaper was allowed to dangle where it had peeled through old age; and there were many places along the hallway and behind doors where plaster had fallen away, leaving gaping holes in the wall.

Stephen's room was apparently little different. It was the magician's lair, the mad professor's laboratory and the messy teenager's study all rolled into one. Among the general detritus and debris, half-finished homework, mugs of undrunk tea, schoolbooks and bits of model aircraft and bizarre gadgets lay in untended heaps. On the sideboard stood electrical devices, the uses of which could only be guessed at, and next to those a rack of test-tubes, their contents neglected and discoloured among the general confusion of odd pieces of wire, paper, glue and metal from half-finished and forgotten projects.

The Hawking family were definitely an eccentric lot. In many ways they were a typically bookish family, but with a streak of originality and social awareness that made them ahead of their time. One contemporary of Hawking's has described them as 'bluestocking'. There were a lot of them; one photograph from the family album includes eighty-eight Hawkings. Stephen's parents did some pretty oddball things. For many years the family car was a London taxi which Frank and Isobel had purchased for £50, but this was later replaced with a brand-new green Ford Consul – the archetypal late-fifties car. There was a good reason for buying it: they had decided to embark on a year-long overland expedition to India, and their old London taxi would never have made it. With the exception of Stephen, who could not interrupt his education, the whole family made the trip to India and back in the green Ford Consul, an astonishingly unusual thing to do in the late 1950s. Needless to say, the vehicle was not in its original pristine condition upon its return.

The Hawkings' journeys outside St Albans were not always so adventurous. Like many families, they kept a caravan on the south coast of England; theirs was near Eastbourne in

Sussex. Unlike other families, however, they owned not a modern version but a brightly coloured gypsy caravan. Most summers, the family spent two or three weeks walking the clifftops and swimming in the bay. Often Stephen's closest friend, John McClenahan, would join them, and the two boys would spend their time flying kites, eating ice-cream and thinking up new ways to tease Stephen's two younger sisters, Mary and Philippa, while generally ignoring his adopted brother Edward, who was only a toddler at the time.

Frank Hawking is significant in Stephen's childhood and adolescence by his absence. He seems to have been a somewhat remote figure who would regularly disappear for several months each year to further his medical research in Africa, sometimes missing the family holidays in Ringstead Bay and leaving the children with Isobel. This routine was so well embedded in the structure of their lives that it was not until her late teens that Stephen's eldest sister, Mary, realized that their family life was at all unusual – she had thought all fathers were like birds that migrated to sunnier climes each year. Whether at home or abroad, Frank Hawking kept meticulous accounts of everything he did in a collection of diaries maintained until the day he died. He also wrote fiction, completing several unpublished novels. One of his literary efforts was written from a woman's viewpoint. Although Isobel respected his efforts when she read it, she believed that it was unsuccessful.

Isobel had an indisputable influence on her eldest son's political ideas. She, like many other English intellectuals of the period, had politically left-of-centre ideas which in her case led to active membership of the St Albans Liberal Association in the 1950s. By then the Liberal Party was only a minor parliamentary force with just a handful of MPs, but at grass-roots level it remained a lively forum for political discussion, often taking the lead, during the 1950s and 1960s, on many issues of the time, including nuclear disarmament and opposition to apartheid. Stephen has never been extreme in his political views, but his interest in politics and left-wing sympathies have never left him.

Stephen and his friends quickly tired of board games and moved on to other hobbies. They built model aircraft and electronic gadgets. The planes rarely flew properly, and Hawking was never as good with his hands as he was with his brain. His model aircraft were usually scruffy constructions of paper and balsa-wood, and far from aerodynamically efficient. With electronics he had similar setbacks, once receiving a 500-volt shock from an old television set he was trying to convert into an amplifier.

In the third and fourth forms, the motley gang of friends began to turn its attention towards the mystical and the religious. Towards the end of 1954, a boy on the periphery of the group, Graham Dow, got religion in a very big way. The evangelist Billy Graham had toured Britain that year, and the young Dow had been greatly influenced by the man. Dow went on to convert Roger Ferneyhaugh, and the enthusiasm spread. Hawking's attitude to this craze is open to debate. Most likely, he stood back from this particular game with a certain amused detachment; this at least is the opinion of his contemporaries. They speak of experiencing a towering intellect, looking on at the reaction of the participants more with fascination than with any feelings of conviction or budding faith.

Michael Church describes how he felt an indefinable intellectual presence when it came to discussing matters vaguely mystical or metaphysical with Stephen. Remembering one encounter, he says:

> I wasn't a scientist and didn't take him remotely seriously until one day when we were messing around in his cluttered, joke-inventor's den. Our talk turned to the meaning of life – a topic I felt pretty hot on at the time – when suddenly I was arrested by an awful realization: he was encouraging me to make a fool of myself, and

watching me as though from a great height. It was a profoundly unnerving moment.[2]

Their interest in Christianity lasted for most of the year. The group of friends met at each other's houses as they had done to play board games. They still drank orange squash, and even played games occasionally, but for most of the time they would hold intense discussions on matters of faith, God and their own feelings. It was a time of inner growth, a struggle to find meaning in the tumble of events and stimuli surrounding them, but it was also an important group activity. One of the group has since intimated that there was an undoubted tinge of schoolboy homosexuality about the whole thing.

This was a difficult time for Stephen. He wanted to be involved, to be part of the group, but the rationalist in him would not, even then, allow his emotions to compromise his intellect. Yet he managed to keep his friends, remain detached and learn a number of social skills which would hold him in good stead for the future. The irony is that at the end of the third year, at the height of the craze, Stephen won the school divinity prize.

After Christianity came the occult. The group began to turn its attention to extrasensory perception (ESP), which at the time was beginning to capture the public imagination. Together and in the privacy of their own dens, they started to conduct experiments during which they would attempt to influence the throw of a die by the power of their minds. Stephen was far more interested in this – it was quantifiable, real experimental work, and there was a chance that the idea could be proved or disproved. It was not simply a matter of faith and hope.

The craze did not last long. With the others, Stephen attended a lecture by a scientist who had made a study of a set of ESP experiments conducted at Duke University in

North Carolina in the late fifties. The lecturer demonstrated that when the experimenters obtained good results the experiments could be shown to be faulty, and whenever the experimental technique was followed correctly, no results were obtained. Hawking's interest turned to contempt. He came to the conclusion that it is only people who have not developed their analytical faculties beyond those of a teenager who believe in such things as ESP.

Meanwhile, at school things carried on pretty much as before. Stephen was poor at all sports with the possible exception of cross-country running, for which his wraith-like physique was perfectly suited. He endured cricket and rugby, but special loathing was reserved for the Combined Cadet Force, the CCF. Like most private boys' schools in Britain, St Albans School maintains a schoolboy army, the original aim of which was to prepare young men for National Service. Each Friday the entire school, with six exceptions, wore military uniform. The exceptions were those whose parents were conscientious objectors. Despite Isobel Hawking's political leanings, Stephen's parents did not object and he took part in the same war games, drills and parades as the others.

For those with little interest in things military, the memories of the CCF are sour – cold winter Fridays in driving rain, clothes drenched through, biting January sleet numbing face and fingers, and the enthusiastic boy-officers yelling orders. Stephen had the rank of Lance Corporal in the Signals, the section into which those with a scientific bent were traditionally placed. By all accounts he hated every minute of it, but it was endured. In some respects the alternative was worse. Those who did not wish to play their part in defending Queen and Country had to run the gauntlet of persuasive tactics. First, the objector was taken to Colonel Pryke, Commander of the CCF. If he did not manage to persuade the dissenter to join, the next line of attack was the Sub-Dean,

Canon Feaver, a formidable gentleman who would subject the boy to a lecture on his moral duty to serve God and the Queen, to play his role in the greater scheme of things. If that were endured, the final test would be to face the headmaster, William Thomas Marsh.

Marsh was one of St Albans' most severe but successful headmasters. He has been described by more than one of Hawking's contemporaries as 'absolutely terrifying'; to cross him was an act of extreme foolishness. If the headmaster failed to convert a conscientious objector, then he must possess tremendous conviction and determination. However, that was only the beginning of the ordeal. Those who did not take part in the CCF were made to dress in fatigues along with everyone else and, instead of playing at soldiers, they were forced to dig a Greek theatre in the school grounds. Marsh was a dedicated Classicist, and he viewed this treatment as fitting ritualistic humiliation. The construction of the Greek theatre continued, come rain or shine, for as long as it took. As the work progressed, Marsh stalked its perimeter in fair weather or surveyed the site from the comfort of a warm room when it was raining or snowing.

Life at school was not always bleak. The whole class often went on school trips to places of academic interest. It was usually the CCF Commander, Colonel Pryke, who was given the responsibility of taking what he referred to as 'a scruffy band of young men' to such places as chemical plants, power stations and museums. He remembers with fondness the occasion when he took Hawking's class to the ICI chemical plant at Billingham in the north of England. Everything seemed to be going well until just after lunch, when one of the scientists who had been showing them around cornered Pryke and said angrily, 'Who the hell have you got here? They're asking me all sorts of bloody awkward questions I can't answer!'

By the time he was fourteen, Stephen knew that he wanted

to make a career out of studying mathematics and it was around this time that his scientific aptitude began to show. He would spend very little time on maths homework and still obtain full marks. As a contemporary recalled, 'He had incredible, instinctive insight. While I would be worrying away at a complicated mathematical solution to a problem, *he just knew the answer* – he didn't have to think about it.'[4] The 'average' bright kid was beginning to reveal a prodigious talent.

One particular example of Stephen's highly developed insight left a lasting impression on John McClenahan. During a sixth-form physics lesson, the teacher posed the question, 'If you have a cup of tea, and you want it with milk and it's far too hot, does it get to a drinkable temperature quicker if you put the milk in as you pour the tea, or should you allow the tea to cool down before adding the milk?' While his contemporaries were struggling with a muddle of concepts to argue the point, Stephen went straight to the heart of the matter and almost instantly announced the correct answer: 'Ah! Milk in last, of course,' and then went on to give a thorough explanation of his reasoning.

He sailed through his Ordinary Level exams, obtaining nine in July 1957 and his tenth, in Latin, a year later, midway through his Advanced Levels. When he sat down to decide on his A Level subjects, parental pressure began to play a part in his plans. He wanted to do Mathematics, Physics and Further Mathematics in preparation for a university course in physics or mathematics. However, Frank Hawking had other plans. He wanted his son to follow him into a career in medicine, for which Stephen would have to study A Level chemistry. After much discussion and argument, Stephen agreed to take Mathematics, Physics and Chemistry A Levels, leaving open the question of university course until the need for a final decision arose a year later.

The sixth form was probably Hawking's happiest time at

St Albans. The boys were allowed greater freedom in their final two years, and they basked a little in the respect they had gained by their success at O Level. In the sixth form, the close group of school-friends began to fragment as their A Level subjects diverged. Those taking arts subjects began, quite naturally, to lose touch with the 'scientists' and different cliques established themselves. Basil King, John McClenahan and Hawking took solely science subjects; the others followed the arts. The scientists gathered others of like mind around them and new groups formed.

In the spring of 1958, Hawking and his friends, including new recruits to the group, Barry Blott and Christopher Fletcher, built a computer called LUCE – the Logical Uniselector Computing Engine. In the 1950s, in Britain only a few university departments and the Ministry of Defence had computers. However, with the help and enthusiasm of a young maths master called Dick Tartar, who had been recruited specifically to generate new ideas and inject some life into the mathematics department, they designed and built a very primitive logic machine.

It took a month to get anything at all out of the machine. The biggest problem, it seems, was not the design or the theoretical side of the project, but simply bad soldering. The guts of the device were recycled parts from an old office telephone exchange, but a vast number of electrical connections were needed to make the device work and the group were forever finding faults in their soldering. Nevertheless, when they did eventually get it to work, it caused considerable excitement in the sixth form. The Mathematical Society write-up in the *Albanian*, the school magazine, sounds as though plucked straight from a time-warp:

It is not unknown for the mathematician to leave his ivory tower and fulfil his original role as a calculator. Thus in 1641 Pascal invented an arithmetical machine – forerunner of the modern computer which specifically replaces tally-stick, abacus or slide-rule as an aid to calculation. Until the happy day when every fourth-former has his pocket Ernie,* we have to be content with logarithm tables. Meanwhile, as a modest start we have LUCE, the St Albans School Logical Uniselector Computing Engine.

This machine answers some useless, though quite complex logical problems. Last term's meetings of the society were devoted to it and proved lively and well attended. With gained experience [the designers] forge ahead with the construction of a digital computer, as yet unchristened, that will actually 'do sums'. (Fourth-formers, take heart!)[5]

Hawking and his friends received their first exposure to the press when the local newspaper, the *Herts Advertiser*, covered the story of 'the schoolboy boffins' building their newfangled machine. And, as promised in the school magazine article, they did go on to make a more sophisticated version of the machine later in the sixth form.

When the present Head of Computing at St Albans School, Nigel Wood-Smith, took over the post many years later, he found a box under one of the tables in the maths room. To him the box appeared to contain nothing more than a pile of old junk, transistors and relays, with 'LUCE' on a name-plate lying discarded atop the tangle of wire and metal. He deposited the entire jumble in the rubbish-bin. It was only many years later that he realized how, unaware of the potential historical significance of things, he had thrown out the computer that Stephen Hawking had built.

* The computer used to select winners of the national lottery.

2

Classical Cosmology

Cosmology is the study of the Universe at large, its beginning, its evolution, and its ultimate fate. In terms of ideas, it is the biggest of big science. Yet in terms of hardware, it is less impressive. True, cosmologists do make use of information about the Universe gleaned from giant telescopes and space probes, and they do sometimes use large computers to carry out their calculations. But the essence of cosmology is still mathematics, which means that cosmological ideas can be expressed in terms of equations written down using pencil and paper. More than any other branch of science, cosmology can be studied by using mind alone. This is just as true today as it was seventy-five years ago, when Albert Einstein developed the general theory of relativity, and thereby invented the science of theoretical cosmology.

When scientists refer to the 'classical' ideas of physics, they are not referring back to the thoughts of the Ancient Greeks. Strictly speaking, classical physics is the physics of Isaac Newton, who laid the foundations of the scientific method for investigating the world, back in the seventeenth century. Newtonian physics reigned supreme until the end of the nineteenth century, when it was overtaken by two revolutions, the first sparked off by Einstein's general theory of relativity, and the second by the quantum theory. The first is the best theory we have of how gravity works; the second explains how everything else in the material world works. Together, these two topics, relativity theory and quantum mechanics, form the twin pillars of modern, twentieth-century science. The Holy Grail of modern physics, sought by many, is a theory that will combine the two into one mathematical package.

But to the modern generation of Grail-seekers in the 1990s, even these twin pillars of physics, in their original form, are old hat. There is another, more colloquial way in which scientists use the term 'classical physics' – essentially, to refer to anything developed by previous generations of researchers, and therefore more than about twenty-five years old. In fact, going back twenty-five years from today does bring us to a landmark event in science: the discovery of pulsars, in 1967, the year Stephen Hawking celebrated his own twenty-fifth birthday. These objects are now known to be neutron stars, the collapsed cores of massive stars that have ended their lives in vast outbursts known as supernova explosions. It was the discovery of pulsars, collapsed objects on the verge of becoming black holes, that revived interest in the extreme implications of Einstein's theory of gravity; and it was the study of black holes that led Hawking to achieve the first successful marriage between quantum theory and relativity.

Typically, though (as we shall see), Hawking was already working on the theory of black holes at least two years before the discovery of pulsars, when only a few mathematicians bothered with such exotic implications of Einstein's equations, and the term 'black hole' itself had not even been used in this connection. Like all his contemporaries, Hawking was brought up, as a scientist, on the classical ideas of Newton and on relativity theory and quantum physics in their original forms. The only way we can appreciate how far the new physics has developed since then, partly with Hawking's aid, is to take a look at those classical ideas ourselves, a gentle workout in the foothills before we head for the dizzy heights. 'Classical cosmology', in the colloquial sense, refers to what was known prior to the revolution triggered by the discovery of pulsars – exactly the stuff that students of Hawking's generation were taught.

Isaac Newton made the Universe an ordered and logical place. He explained the behaviour of the material world in terms of fundamental laws which were seen to be built into the fabric of the Universe. The most famous example is his law of gravity. The orbits of the planets around the Sun had remained a deep mystery before Newton's day, but he explained them by a law of gravity which says that a planet at a certain distance from the Sun feels a certain force, tugging on it, proportional to one over the square of the distance to the Sun – what is known as an inverse square law. In other words, if the planet is magically moved out to twice as far from the Sun, it will feel one-quarter of the force; if it is put three times as far away, it will feel one-ninth of the force; and so on. As a planet in a stable orbit moves through space at its own speed, this inward force exactly balances the tendency of the planet to fly off into space. Moreover, Newton realized, the same inverse square law explains the fall of an apple from a tree and the orbit of the Moon about the Earth, and even the ebb and flow of the tides. It is a *universal* law.

Newton also explained the way in which objects respond to forces other than gravity. Here on Earth, when we push something it moves, but only as long as we keep pushing it. Any moving object on Earth experiences a force, called friction, which opposes its motion. Stop pushing, and friction will bring the object to a halt. Without friction, though (like the planets in space, or the atoms that everyday things are composed of), according to Newton, an object will keep moving in a straight line at a steady speed until a force is applied to it. Then, as long as the force continues to operate, the object will accelerate, changing its direction, or its speed, or both. The lighter the object, or the stronger the force, the greater the acceleration that results. Take away the force, however, and once again the object moves at a steady speed in a straight line, but at the new velocity that has built up during the time it was accelerating.

When you push something, it pushes back, and the action and reaction are equal and opposite. This is how a rocket works – it throws material out from its exhaust in one direction, and the reaction pushes the rocket along in the opposite direction. This last law is familiar these days from the snooker table, where balls collide and rebound off each other in a very 'Newtonian' manner. And that is very much the image of the world that comes out of Newtonian mechanics – an image of balls (or atoms) colliding and rebounding, or of stars and planets moving under the influence of gravity, in an exactly regular and predictable manner.

All these ideas were encapsulated in Newton's master-work, the *Principia*, published in 1687 (usually referred to simply by the short version of its Latin title; the full English title of Newton's great work is *Mathematical Principles of Natural Philosophy*). The view Newton gave us of the world is sometimes referred to as the 'clockwork universe'. If the Universe is made up of material objects interacting with each other through forces that obey truly universal laws, and if rules like that of action and reaction apply precisely throughout the Universe, then the Universe can be regarded as a gigantic machine, a kind of cosmic clockwork, which will follow an utterly predictable path forever once it has been set in motion.

This raises all kinds of puzzles, deeply worrying to philosophers and theologians alike. The heart of the problem is the question of free will. In such a clockwork universe, is everything predetermined, including all aspects of human behaviour? Was it pre-ordained, built into the laws of physics, that a collection of atoms known as Isaac Newton would write a book known as the *Principia* that would be published in 1687? And if the Universe can be likened to a cosmic clockwork machine, who wound up the clockwork and set it going?

Even within the established framework of religious belief in seventeenth-century Europe, these were disturbing questions, since although it might seem reasonable to say that the clockwork could have been wound up and set in motion by God, the traditional Christian view sees human beings as having free will, so that they can choose to follow the teachings of Christ or not, as they wish. The notion that sinners might actually have no freedom of choice concerning their actions, but were sinning in obedience to inflexible laws, following a path to eternal damnation actually laid out by God in the beginning, simply could not be fitted into the established Christian world-view.

Strangely, though, in Newton's day, and down into the twentieth century, science did not really contemplate the notion of a beginning to the Universe at all. The Universe at large was perceived as eternal and unchanging, with 'fixed' stars hanging in space. The biblical story of the Creation, still widely accepted in the seventeenth century by scientists as well as ordinary people, was thought of as applying only to our planet, the Earth, or perhaps to the Sun's family, the Solar System, but not to the whole Universe.

Newton believed (incorrectly, as it turns out) that the fixed stars could stay as they were in space forever if the Universe were infinitely big, because the force of gravity tugging on each individual star would then be the same in all directions. In fact, such a situation is highly unstable. The slightest deviation from a perfectly uniform distribution of stars will produce an overall pull in one direction or another, making the stars start to move. As soon as a star moves towards any source of gravitational force, the distance to the source decreases, so the force gets stronger, in line with Newton's inverse square law. So, once the stars have started to move, the force causing the non-uniformity gets bigger, and they keep on moving at an accelerating rate. A static universe will soon start to collapse under the pull of gravity. But that became clear only after Einstein had developed a new theory of gravity – a theory, moreover, which contained within itself a prediction that the Universe would certainly not be static, and might actually be, not collapsing, but *expanding*.

Like Newton, Albert Einstein made many contributions to science. Also like Newton, his master-work was his theory of gravity, the general theory of relativity. It is some measure of just how important this theory is to the modern understanding of the Universe that even Einstein's special theory of relativity, the one that leads to the famous equation $E = mc^2$, is by comparison a relatively minor piece of work. Nevertheless, the special theory, which was published in 1905, contributed a key ingredient to the new understanding of the Universe. Before we move on to this, though, we should at least give a brief outline of the main features of the special theory.

Einstein developed the special theory of relativity in response to a puzzle that had emerged from nineteenth-century science. The great Scottish physicist, James Clerk Maxwell, had found the equations that describe the behaviour of electromagnetic waves. Maxwell's equations were soon developed to explain the behaviour of radio waves, which were discovered in 1888. But Maxwell had found that the equations automatically gave him a particular speed,* which is identified as the speed at which electromagnetic waves travel. The unique speed that came out of Maxwell's equations turned out to be exactly the speed of light, which physicists had already measured by that time. This revealed that light must be a form of electromagnetic wave, like radio waves but with shorter wavelength (that is, higher frequency). And it

* Strictly speaking, it is a velocity – a quantity which specifies speed and direction. For our purposes, it is easier to refer to velocities as speeds.

also meant, according to those equations, that light (as well as other forms of electromagnetic radiation, including radio waves) always travels at the same speed.

This is not what we expect from our everyday experience of how things move. If I stand still, and toss a ball to you gently, it is easy for you to catch the ball. If I am driven towards you at 60 miles an hour in a car, and toss the ball equally gently out of the window, it hurtles towards you at 60 miles an hour plus the speed of the toss. You would, rightly, be dumbfounded if the ball tossed gently out of the car window reached you travelling only at the gentle speed of the toss, without the speed of the car being added in, yet that is exactly what happens with light pulses. Equally, if one vehicle travelling at 50 miles an hour along a straight road is overtaken by another travelling at 60 miles an hour, the second vehicle is moving at 10 miles an hour relative to the first one. Speed, in other words, is relative. And yet, if you are overtaken by a light pulse, and measure its speed as it goes past, you will find it has the same speed you would measure for a light pulse going past you when you are standing still.

Nobody knew this until the end of the nineteenth century. Scientists had assumed that light behaved in the same way, as far as adding and subtracting velocities is concerned, as objects like balls being thrown from one person to another. And they explained the 'constancy' of the speed of light in Maxwell's equations by saying that the equations applied to some 'absolute space', a fundamental reference frame for the entire Universe.

According to this view, space itself defined the framework against which things should be measured – absolute space, through which the Earth, the Sun, light and everything else moved. This absolute space was also sometimes called the 'aether', and conceived of as a substance through which electromagnetic waves moved, like water waves moving over the sea. The snag was, when experimenters tried to measure changes in the velocity of light caused by the motion of the Earth through absolute space (or 'relative to the aether'), none could be found.

Because the Earth moves round the Sun in a roughly circular orbit, it should be moving at different speeds relative to absolute space at different times of the year. It's like swimming in a circle in a fast-flowing river. Sometimes the Earth will be 'swimming with the aether', sometimes across the aether, and sometimes against the flow. If light always travels at the same speed relative to absolute space, then, common sense tells us, this ought to show up in the form of seasonal changes in the speed of light measured from the Earth. It does not.

Einstein resolved the dilemma with his special theory. This says that *all* frames of reference are equally valid, and that there is no absolute reference frame. Anybody who moves at a constant velocity through space is entitled to regard themselves as stationary. They will find that moving objects in their frame of reference obey Newton's laws, while electromagnetic radiation obeys Maxwell's equations and the speed of light is always measured to be the value that comes out of those equations, denoted by the letter c. Furthermore, anybody who is moving at a constant speed relative to the first person (the first observer, in physicists' jargon) will also be entitled to say that they are at rest, and will find that objects in their laboratory obey Newton's laws, while measurements always give the speed of light as c. Even if one observer is moving towards the other observer at half the speed of light, and sends a torch beam out ahead, the second observer will not measure the speed of the light from the torch as $1.5c$: it will still be c!

Starting out from the observed fact that the speed of light is a constant, the same whichever way the Earth is moving through space, Einstein found a mathematical package to describe the behaviour of material objects in reference frames that move with constant velocities relative to one another – so-called 'inertial' frames of reference. Provided the velocities are small compared with the speed of light, these equations give exactly the same 'answers' as Newtonian mechanics. But when the velocities begin to become an appreciable fraction of the speed of light, strange things happen.

Two velocities, for example, can *never* add up to give a relative velocity greater than c. An observer may see two other observers approaching each other on a head-on collision course, each travelling at a speed of $0.9c$ in the first observer's reference frame; but measurements carried out by either of those two fast-moving observers will always show that the other one is travelling at a speed less than c, but bigger (in this case) than $0.9c$.

The reason why velocities add up in this strange way has to do with the way both space and time are warped at high velocities. In order to account for the constancy of the speed of light, Einstein had to accept that moving clocks run more slowly than stationary clocks, and that moving objects shrink in the direction of their motion. The equations also tell us that moving objects increase in mass the faster they go.

Strange and wonderful though all these things are, they are only peripheral to the story of modern cosmology and to the search for links between quantum physics and gravity. We stress, however, that they are not wild ideas, in the sense that we sometimes dismiss crazy notions as 'just a theory' in everyday language. To scientists, a theory is an idea that has been tried and tested by experiments, and has passed every test. The special theory of relativity is no exception to this rule. All the strange notions implicit in the theory – the constancy of the speed of light, the stretching of time and shrinking of length for moving objects, the increase in mass of a moving object – have been measured and confirmed to great precision in very many experiments. Particle accelerators – 'atom smashing' machines like those at CERN, the

European Centre for Nuclear Research, in Geneva – simply would not work if the theory were not a good one, since they have been designed and built around Einstein's equations. The special theory of relativity as a description of the high-speed world is as securely founded in solid experimental facts as is Newtonian mechanics as a description of the everyday world; the only reason it conflicts with our common sense is that in everyday life we are not used to the kind of high-speed travel required for the effects to show up. After all, the speed of light, *c*, is 300,000 kilometres a second (186,000 miles a second), and the relativistic effects can be safely ignored for any speeds less than about 10 per cent of this – that is, for speeds less than a mere 30,000 kilometres a second.

In essence, the special theory is the result of a marriage of Newton's equations of motion with Maxwell's equations describing radiation. It was very much a child of its time, and if Einstein hadn't come up with the theory in 1905, one of his contemporaries would surely have done so within the next few years. Without Einstein's special genius, though, it might have been a generation or more before anyone realized the importance of a far deeper insight buried within the special theory.

This key ingredient, to which we have already alluded, was the fruit of another marriage – the union of space and time. In everyday life, space and time seem to be quite different things. Space extends around us in three dimensions (up and down, left and right, forwards and backwards). We can see where things are located in space, and travel through it more or less at will. Time, although we all know what it is, is almost impossible to describe. In a sense, it does have a direction (from past to future), but we can look neither into the future nor into the past, and we certainly cannot move through time at will. Yet the great universal constant, *c*, is a speed, and speed is a measure that relates space and time.

Speeds are always in the form of miles per hour, or centimetres per second, or any other unit of length per unit of time. You cannot have one without the other when you are talking about speed. So the fact that the fundamental constant is a velocity must be telling us something significant about the Universe. But what?

If you multiply a speed by a time, you get a length. And if you do this in the right way (by multiplying intervals of time by the speed of light, c) you can combine measures of length (space) with measures of time in the same set of equations. The set of equations that combines space and time in this way consists of the equations of the special theory of relativity that describe time dilation and length contraction, and lead to the prediction that a mass m is equivalent to an energy E as described by the formula $E = mc^2$. Instead of thinking about space and time as two separate entities, as long ago as 1905 Einstein was telling physicists that they should be thinking about them as different aspects of a single, unified whole – spacetime. But this spacetime, the special theory also said, was not fixed and permanent like the absolute space or absolute time of Newtonian physics – it could be stretched or squeezed. And therein lay the clue to the next great step forward.

Einstein used to say that the inspiration for his general theory of relativity (which is, above all, a theory of gravity) came from the realization that a person inside a falling lift whose cable had snapped would not feel gravity at all. We can picture exactly what he meant, because we have now seen film of astronauts orbiting the Earth in spacecraft. Such an orbiting spacecraft is not 'outside' the influence of the Earth's gravity; indeed, it is held in orbit by gravity. But the spacecraft and everything in it is falling around the Earth with the same acceleration, so the astronauts have no weight and float within their capsule. For them, it is as if gravity does not exist, a phenomenon known as free fall. But Einstein had never seen any of this, and had to picture the situation in a freely falling lift in his imagination. It is as if the acceleration of the falling lift, speeding up with every second that passes, precisely cancels out the influence of gravity. For that to be possible, gravity and acceleration must be exactly equivalent to one another.

The way this led Einstein to develop a theory of gravity was through considering the implications for a beam of light, the universal measuring tool of special relativity. Imagine shining a torch horizontally across the lift from one side to the other. In the freely falling lift, objects obey Newton's laws: they move in straight lines, from the point of view of an observer in the lift, bounce off each other with action and reaction equal and opposite, and so on. And, crucially, from the point of view of the observer in the lift, light travels in straight lines.

But how do things look to an observer standing on the ground, watching the lift fall? The light would appear to follow a track that always stays exactly the same distance below the roof of the lift. But in the time it takes the light to cross the lift, the lift has accelerated downwards, and the light in the beam must have done the same. In order for the light to stay the same distance below the roof all the way across, the light pulse must follow a curved path as seen from outside the lift. In other words, a light beam must be bent by the effect of gravity.

Einstein explained this in terms of bent spacetime. He suggested that the presence of matter in space distorts the spacetime around it, so that objects moving through the distorted spacetime are deflected, just as if they were being tugged in ordinary 'flat' space by a force inversely proportional to the square of the distance. Having thought up the idea, Einstein then developed a set of equations to describe all this. The task took him ten years. When he had finished, Newton's famous inverse square law re-emerged from Einstein's new theory of gravity; but general relativity went far beyond Newton's theory, because it also offered an all-embracing theory of the whole Universe. The general theory describes all of spacetime, and therefore all of space and all of time. (There is a neat way to remember how it works. Matter tells spacetime how to bend; bends in spacetime tell matter how to move. But, the equations also insisted, spacetime itself can also move, in its own fashion.)

The general theory was completed in 1915 and published in 1916. Among other things, it predicted that beams of light from distant stars, passing close by the Sun, would be bent as they moved through spacetime distorted by the Sun's mass. This would shift the apparent positions of those stars in the sky – and the shift might actually be seen, and photographed, during a total eclipse, when the Sun's blinding light is blotted out. Just such an eclipse took place in 1919; the photographs were taken, and showed exactly the effect Einstein had predicted. Bent spacetime was real: the general theory of relativity was correct.

But the equations developed by Einstein to describe the distortion of spacetime by the presence of matter, the very equations that were so triumphantly vindicated by the eclipse observations, contained a baffling feature that even Einstein could not comprehend. The equations insisted that the spacetime in which the material Universe is embedded could not be static. It must be either expanding, or contracting.

Exasperated, Einstein added another term to his equations, for the sole purpose of holding spacetime still. Even at the beginning of the 1920s, he still shared (along with all his contemporaries) the Newtonian idea of a static Universe. But within ten years, observations made by Edwin Hubble with a new and powerful telescope on a mountain-top in California had shown that the Universe *is* expanding.

The stars in the sky are not moving farther apart from one another. They belong to a huge system, the Milky Way

Galaxy, which contains about a hundred billion stars and is like an island in space. In the 1920s, astronomers discovered with the aid of new telescopes that there are many other galaxies beyond the Milky Way, many of them containing hundreds of billions of stars like our Sun. And it is the galaxies, not individual stars, that are receding from one another, being carried farther apart as the space in which they are embedded expands.

If anything, this was an even more extraordinary and impressive prediction of the general theory than the bending of light detectable during an eclipse. The equations had predicted something that even Einstein at first refused to believe, but which observations later showed to be correct. The impact on scientists' perception of the world was shattering. The Universe was not static, after all, but evolving; Einstein later described his attempt to fiddle the equations to hold the Universe still as 'the greatest blunder of my life'. Even at the end of the 1920s, the observations and the theory agreed that the Universe is expanding. And if galaxies are getting farther apart, that means that long ago they must have been closer together. How close could they ever have been? What happened in the time when galaxies must have been touching one another, and before then?

The idea that the Universe was born in a superdense, superhot fireball known as the Big Bang is now a cornerstone of science, but it took time – over fifty years – for this theory to become developed. Just at the time astronomers were finding evidence for the universal expansion, transforming the scientific image of the Universe at large, their physicist colleagues were developing the quantum theory, transforming our understanding of the very small. Attention focused chiefly on the development of the quantum theory over the next few decades, with relativity and cosmology (the study of the Universe at large) becoming an exotic branch of science investigated only by a few specialist mathematicians. The union of large and small still lay far in the future, even at the end of the 1920s.

As the nineteenth century gave way to the twentieth, physicists were forced to revise their notions about the nature of light. This initially modest readjustment of their world-view grew, like an avalanche triggered by a snowball rolling down a hill, to become a revolution that engulfed the whole of physics – the quantum revolution.

The first step was the realization that electromagnetic energy cannot always be treated simply as a wave passing through space. In some circumstances, a beam of light, for example, will behave more like a stream of tiny particles (now called photons). One of the people instrumental in establishing this 'wave–particle duality' of light was Einstein, who in 1905 showed how the way in which electrons are knocked out of the atoms in a metal surface by electromagnetic radiation (the photoelectric effect) can be explained neatly in terms of photons, not in terms of a pure wave of electromagnetic energy. (It was for this work, not his two theories of relativity, that Einstein received his Nobel prize.)

This wave–particle duality changes our whole view of the nature of light. We are used to thinking of momentum as a property to do with the mass of a particle and its speed (or, more correctly, its velocity). If two objects are moving at the same speed, then the heavier one carries more momentum, and will be harder to stop. A photon does not have mass, and at first sight you might think this means it has no momentum either. But, remember, Einstein discovered that mass and energy are equivalent to one another, and light certainly does carry energy – indeed, a beam of light is a beam of pure energy. So photons do have momentum, related to their energy, even though they have no mass and cannot change their speed. A change in the momentum of a photon means that it has changed the amount of energy it carries, not its velocity; and a change in the energy of a photon means a change in its wavelength.

When Einstein put all of this together, it implied that the momentum of a photon multiplied by the wavelength of the associated wave always gives the same number, now known as Planck's constant in honour of Max Planck, another of the quantum pioneers. Planck's constant (usually denoted by the letter *h*) soon turned out to be one of the most fundamental numbers in physics, ranking alongside the speed of light, *c*. It cropped up, for example, in the equations developed in the early decades of the twentieth century to describe how electrons are held in orbit around atoms. But although the strange duality of light niggled, the cat was only really set among the pigeons in the 1920s when a French scientist, Louis de Broglie, suggested using the wave–particle equation in reverse. Instead of taking a wavelength (for light) and using this to calculate the momentum of an associated particle (the photon), why not take the momentum of a particle (such as an electron) and use it to calculate the length of an associated wave?

Fired by this suggestion, experimenters soon carried out tests which showed that, under the right circumstances, electrons do indeed behave like waves. In the quantum world (the world of the very small, on the scale of atoms and below), particles and waves are simply twin facets of *all* entities. Waves can behave like particles; particles can behave like waves. A term was even coined to describe these quantum entities – 'wavicles'. The dual description of particles as waves and waves as particles turned out to be the key to unlocking the secrets of the quantum world, leading to the development of a satisfactory theory to account for the behaviour of atoms, particles and light. But at the core of that theory lay a deep mystery.

Because all quantum entities have a wave aspect, they

cannot be pinned down precisely to a definite location in space. By their very nature, waves are spread-out things. So we cannot be certain where, precisely, an electron is – and uncertainty, it turns out, is an integral feature of the quantum world. The German physicist Werner Heisenberg established in the 1920s that all observable quantities are subject, on the quantum scale, to random variations in their size, with the magnitude of these variations determined by Planck's constant. This is Heisenberg's famous 'uncertainty principle'. It means that we can *never* make a precise determination of all the properties of an object like an electron: all we can do is to assign probabilities, determined in a very accurate way from the equations of quantum mechanics, to the likelihood that, for example, the electron is in a certain place at a certain time.

Furthermore, the uncertain, probabilistic nature of the quantum world means that if two identical wavicles are treated in an identical fashion (perhaps by undergoing a collision with another type of wavicle), they will not necessarily respond in identical fashions. That is, the outcome of experiments is also uncertain, at the quantum level, and can be predicted only in terms of probabilities. Electrons and atoms are not like tiny snooker balls bouncing around in accordance with Newton's laws.

None of this shows up on the scale of our everyday lives, where objects such as snooker balls do bounce off each other in a predictable, deterministic fashion, in line with Newton's laws. The reason is that Planck's constant is incredibly small: in standard units used by physicists, it is a mere 6×10^{-34} (a decimal point followed by 33 zeros and a 6) of a joule-second. And a joule is indeed a sensible sort of unit in everyday life – a 60-watt light bulb radiates 60 joules of energy every second. For everyday objects like snooker balls, or ourselves, the small size of Planck's constant means that the wave associated with the object has a comparably small wavelength, and can be ignored. But even a snooker ball, or yourself, *does* have an associated quantum wave, even though it is only for tiny objects like electrons, with tiny amounts of momentum, that you get a wave big enough to interfere with the way objects interact.

It all sounds very obscure, something we can safely leave the physicists to worry about while we get on with our everyday lives. To a large extent, that is true, although it is worth realizing that the physics behind how computers or TV sets work depends on an understanding of the quantum behaviour of electrons. Laser beams, also, can be understood only in terms of quantum physics, and every compact disc player uses a laser beam to scan the disc and 'read' the music. So quantum physics actually does impinge on our everyday lives, even if we do not need to be a quantum mechanic to make a TV set or a hi-fi system work. But there is something much more important to our everyday lives inherent in quantum physics. By introducing uncertainty and probability into the equations, quantum physics does away once and for all with the predictive clockwork of Newtonian determinism. If the Universe operates, at the deepest level, in a genuinely unpredictable and indeterministic way, then we are given back our free will, and we can after all make our own decisions and our own mistakes.

At the beginning of the 1960s, the two great pillars of physics stood in splendid separation. General relativity explained the behaviour of the cosmos at large, and suggested that the Universe must have expanded from a superdense state, colloquially known as the Big Bang. Quantum physics explained how atoms and molecules work, and gave an insight into the nature of light and other forms of radiation. One young physicist, taking his first degree at Oxford University, would have been given a thorough grounding in both great theories. But he would hardly have suspected that over the next thirty years he would play a key role in bringing the two theories together, providing insight into how they might be unified into one grand theory that would explain *everything*, from the Big Bang to the atoms we are made of.

3

Going Up

The year 1959 started with a bang: 2 January saw the thirty-two-year-old Fidel Castro sweeping to power in Cuba; a month later Buddy Holly died in a plane crash and Indira Gandhi became the leader of India's ruling Congress Party. By the spring, the world's first hovercraft was under construction on the Isle of Wight, two rhesus monkeys had become the first primates in space, and the writer Raymond Chandler had died aged seventy. Meanwhile, in a small city in Hertfordshire a seventeen-year-old schoolboy called Stephen Hawking was getting ready for the Oxford entrance examination in a large, cluttered bedroom in his parents' rambling Edwardian house.

Obtaining a place at Oxford University was no easy task. A potential candidate had two alternatives – an entrance examination taken in the upper sixth, before A Levels, or the same examination taken in the seventh term, provided very high A Level grades had been obtained. The former route meant that a successful candidate could go straight to Oxford after the summer vacation; the latter necessitated waiting until the following October to go up.

Stephen and his father settled on the first alternative, and he was entered for the examination towards the end of his final year at St Albans School. The intention from the start was that he was going for a scholarship, the highest award offered by the University. The award provided a number of titular privileges and, more important, a percentage of the cost of putting a student through Oxford was paid by the University. A student failing to obtain a scholarship could be awarded an exhibition, which was less prestigious and brought with it a smaller contribution to the costs of education. Last, a candidate could be offered a place at the University but with no financial assistance at all and the student was then known as a 'commoner'.

Over the previous year, father and son had engaged in endless arguments over the choice of university course. Stephen insisted that he wanted to read mathematics and physics, a course then known as Natural Science. His father was unconvinced; he believed there were no jobs in mathematics apart from teaching. Stephen knew what he wanted to do and won the argument; medicine had little appeal for him. As he says himself:

> My father would have liked me to do medicine. However, I felt that biology was too descriptive, and not sufficiently fundamental. Maybe I would have felt differently if I had been aware of molecular biology, but that was not generally known about at the time.[1]

Frank Hawking lost the argument over Stephen's choice of degree course, but he was determined to see his son obtain a place at his old college, University College, Oxford. However, it is clear that Dr Hawking was not, even at this stage, fully convinced of Stephen's ability, and believed that he had to pull strings to get him in. He evidently decided to take the initiative. Just before the scheduled entrance examination in the Easter vacation, he arranged to take Stephen to meet his prospective tutor at University College, Dr Robert Berman. As Berman himself recalls, the sort of pressure Hawking Senior was applying would usually have put him off the candidate immediately. However, Stephen sat the exam and did so extraordinarily well that Berman and University College soon warmed to him.

The entrance examination was pretty tough. It was spread over two days and consisted of five papers in all, each of which was two and a half hours long. These included two physics and two maths papers, followed by a general paper which tested candidates on their general knowledge and awareness of current affairs and world issues. A typical question would have been something like 'Discuss the possible short-term global consequences of Fidel Castro's takeover of Cuba.' It was doubted at the time whether a seventeen-year-old should be expected to have very strong opinions on such matters at all, and some at the University even doubted the desirability of such opinions. Dr Berman, for one, would have been more impressed, he has said, by Hawking's knowledge of the England cricket team than his views on contemporary politics.

After twelve and a half hours of theory examinations and a physics practical paper came the interviews. First there was a general interview at which the candidates were grilled by the Master, Dean, Senior Tutor and Fellows of the subject. These took place in the Senior Common Room. The prospective students were led in individually to face stern appraisal by the panel, and were expected to give intelligent answers to a series of obtuse questions. The purpose of this, like that of a job interview, was to find out a little more about the character of the candidate. Following the general interview a specialist interview was held in Dr Berman's office, during which Hawking was questioned on his knowledge of physics.

The interviews and examinations over, the candidates returned to their various schools around the country to await the results and get on with their A Levels. Meanwhile, the tutors marked the papers and conferred on the matter. If University College wanted Hawking, they had first choice of offering him a scholarship because he had placed them at the top of his list in his application. If they decided they did not want to award him a scholarship or an exhibition, then other

Oxford colleges on his list could take up the option. If no one wanted to give him an award, then the choice went back to University College to take him as a commoner if they wished.

Ten days passed before he heard anything from them. Then came the invitation to return for another interview. This was a promising step forward. It meant that they were taking his application seriously and that he had a very good chance of obtaining a place. Little did he know that he had scored around ninety-five per cent in both his physics papers, with only slightly lower percentages in the others. A few days after the second interview the all-important letter fell on to the Hawkings' doormat. University College was offering him a scholarship. He was invited to enrol at Oxford University the following October, the only condition being that he obtain two A Level passes in the summer.

It has often been said that there is a certain light in Oxford, a wonderful interplay between sunlight and sandstone which, like the comparably beautiful cities of Italy and Germany, has inspired the work of poets and painters down the centuries. The city centre is totally dominated by the presence of the University – a ubiquitous thing, without nerve-centre or organized structure. The colleges are to be found in a random scattering with the rest of the town weaving around it. The architecture displays as little organization as the geography, dating from medieval times to the late twentieth century. On summer days, with the sunlight strong against the stonework and the river dotted with punts, their navigators sweeping a pole into the sparkling water and those on the grassy banks lifting a glass of champagne to their lips, it can, if you let it, seem like an earthly paradise in freeze-frame.

In the late fifties and early sixties, Oxford, as a microcosm of British society, was on the brink of great change. When Hawking arrived in The High on his first October Thursday as an undergraduate, the University had in many respects changed little since his father's time or, indeed, for the past few hundred years. University discipline had relaxed somewhat since the end of the war. Before then, students had been forbidden to enter the city's pubs and could, if caught, be expelled from them by the University Police, known as the Bulldogs. Women were not allowed in male students' rooms without written permission from the Dean, who would specify strict time limitations and conditions in a letter sent to the Head Porter, who would then rigorously uphold the Dean's instructions. All this changed when servicemen returning from the war entered the university either as freshmen or to restart courses interrupted by the fighting. Naturally, they were unwilling to accept such draconian restrictions and gradually the rules were relaxed.

Students vied for rooms in college, but Hawking was lucky in that, being a Scholar, he took priority and managed to keep his room in hall throughout his three years at Oxford.

Most Oxford colleges are built in the form of a number of quads, each with a lawn at the centre and paths around and across the grass. From the quads staircases lead off into the buildings, and the students' rooms are on a number of levels up to the top of each staircase. Students in college had their rooms cleaned for them and their general domestic duties handled by college servants, or 'scouts'. Scouts were also responsible for making sure that hung-over young men and the occasional young woman made it to breakfast between the regulation times of 8.00 to 8.15, so as to avoid facing a locked dining-hall door. Scouts addressed the students as 'Sir', or as 'Mister Such-and-Such' if attempting to inject a note of disdain into their voices. They in turn were addressed by their surnames, in true master–servant fashion.

The intake at Oxford was still largely male, and from the country's private schools, and the majority of those were from the top ten, including Eton, Harrow, Rugby and Westminster.

The number of students from middle-class and working-class backgrounds was beginning to increase, but in many respects the class system took on a greater refinement and a sharper profile at Oxford University. There were definite lines of demarcation. The friendships and relationships capable of crossing those invisible boundaries were still amazingly few. The twain very rarely met.

In one camp were the elite, the children of the aristocracy and heirs to 'old money', the Sebastian Flytes of this world; they made up a substantial proportion of students at Christchurch and, to a lesser extent, Balliol. The privileged spent their often considerable allowances largely on entertaining their chums from school who had gone up with them or friends who had chosen to go to 'the other place', Cambridge. They looked upon those from minor private schools such as St Albans as a lesser breed, lumping them in with the lowest of the low – ex-grammar-school boys. Despite literature's tendency towards exaggeration, it was still all very *Brideshead Revisited.* On the other side of the divide, 'the Northern chemists' and the 'grammar-school oiks' made do on their scholarships and grants, forfeiting quails' eggs and champagne for pork pies and beer.

The two groups looked surprisingly similar in many respects. In the late fifties, baggy trousers and tweed jackets were in fashion for academic young men whatever their background. The difference was that for a privileged few the jackets came from Savile Row and the baggy turn-ups from Harrods. At the college balls held each summer, the female companion of an Old Harrovian or Etonian would, in all probability, be the daughter of a baron or a duke, wrapped in the best silk. Meanwhile, at the same functions, the middle classes gathered together with others of their own kind, sipping at rarely sampled champagne.

A simple point of reference illustrates the changes about to hit Oxford soon after Hawking went up, encapsulated by one of his contemporaries. 'When we arrived in Oxford,' he said, 'anybody who was anybody rowed and never wore jeans. When we left, anybody who was anybody never rowed and did wear jeans.'

Changes were afoot everywhere. Beat poetry from San Francisco was beginning to have an influence. The Labour Party was growing in popularity. The old values, the class system in particular, were beginning to look anachronistic, at least among the intelligentsia. There was no desire to 'storm the citadel' (that would come a decade later and in a different city), but the *Zeitgeist* was definitely on the move. When it came down to it, Hawking's type of person found the whole infrastructure of Oxford as a microcosm faintly amusing, an ethos which would, in peculiarly British fashion, lead to *Beyond the Fringe* and *Monty Python* rather than blood in Parisian gutters.

Despite its many charms, Hawking's first year at Oxford was, by all accounts, a pretty miserable time for him. Very few of his school contemporaries and none of his close friends from St Albans had gone up the same year. In 1960 Michael Church arrived, and John McClenahan went to Cambridge. Many students in Hawking's year had completed National Service before going up, and were consequently a couple of years older than he. (He had avoided the call-up himself by only a few months when it was scrapped by Harold Macmillan's government.)

Work was a bore. He had very little difficulty solving any of the physics or mathematics questions his tutors set him, and he went into a downward spiral of bothering very little and finding meagre satisfaction in easy victories. The system at Oxford made it easy for someone like Hawking to slide into apathy. Students were expected to attend a number of lectures each week and a weekly tutorial in which problems set during the previous tutorial were gone through. Apart from these commitments, students were left largely to their own devices.

On top of this freedom, the examination structure was loose and eminently open to exploitation if you were of Hawking's calibre. The only crucial examinations were set by the University, as opposed to the college, and took place at the end of the first year and again in the final year. The degree was awarded solely on the student's performance in Finals. There were also college exams set at the beginning of each new term to test the students on both the previous term's work and their personal studies during the vacation. These were called Collections and marked by the students' own tutors. As Hawking relates:

> The prevailing attitude at Oxford at that time was very anti-work. You were supposed either to be brilliant without effort or to accept your limitations and get a fourth-class degree. To work hard to get a better class of degree was regarded as the mark of a grey man, the worst epithet in the Oxford vocabulary.[2]

Hawking knew that he was in the former category, and determined to live up to the image. During his first year he attended only mathematics lectures and tutorials, and completed college exams solely in mathematics. As his tutor now freely admits, the physics course at the time was little more than a repetition of A Level work and of limited use to the Hawkings of this world.

There has arisen a veritable folk-tradition of anecdotes about his intuitive understanding of the subject at university, stories reminiscent of the early prowess of the boy-Mozart. One of his contemporaries who shared tutorials with Hawking recalls an incident which left a lasting impression on him. They had been set some problems by the tutor to bring to the next tutorial. No one in the group could do them except Stephen. The tutor asked to see his work and was immensely impressed with his proof of a particularly difficult theorem and, complimenting him on the achievement, handed back the paper. Without the slightest hint of arrogance Hawking took back his work, screwed it into a ball and lobbed it into the wastepaper bin in the far corner of the room. Another member of the tutorial group said later, 'If I had been able to prove that in a year, I would have kept it!'

Another story tells of the time the four members of his tutorial group were set a collection of problems for the following week. On the morning the questions were due in, the other three came across Hawking in the common-room slouched in an armchair reading a science-fiction novel.

'How have you got on with the problems, Steve?' one of them asked.

'Oh, I haven't tried them,' Hawking replied.

'Well, you'd better get on with it,' said his friend. 'The three of us have been working on them together for the past week and we've only managed to get one of them done.'

Later that day the three of them encountered Hawking walking to the tutorial and enquired how he had got on with the problems. 'Oh,' he said, 'I only had time to do nine of them.'

Hawking kept very few notes and possessed only a handful of textbooks. In fact, he was so far ahead of the field that he had become distrustful of many standard textbooks. A further anecdote describes the time one of his tutors, a junior research fellow called Patrick Sandars, set the class some problems from a book. Hawking turned up to the following tutorial having failed to complete any of the questions. When asked why, he spent the next twenty minutes pointing out all the errors in the textbook.

Despite his lackadaisical attitude to things academic, he still managed to maintain a healthy relationship with his tutor, Dr Berman. He would occasionally go for tea at the Bermans' home in Banbury Road. In the summer they would hold parties on the back lawn, at which they would eat strawberries and play croquet. Dr Berman's wife, Maureen, took a particular liking to the rather eccentric young student whom her husband rated so highly as a physicist. Hawking would often arrive early for tea to ask her advice on what good books he should buy, and she guided him through a highbrow literary diet to supplement the physics texts he would occasionally read.

His lack of effort hardly seemed to hold back his progress in physics. As an award student, he had to enter for the University Physics Prize at the end of the second year, for which all the other physicists in his year entered. With the minimum of effort he won the top award and received a Blackwell's book token for £50.

Maintaining his academic position in college and staying on the right side of Dr Berman was one thing, but coping with the increasing boredom of it all was quite another, and at this time he may have nosedived into depression. Fortunately, in the second year he discovered an interest which would help him find some sort of stability. He took up rowing. Rowing has a long tradition at both Oxford and Cambridge, dating back centuries. Each year the Boat Race between the two universities highlights the skills of the best oarsmen who spend the rest of the year in intercollegiate races and training.

Rowing is a very physical activity, and is taken terribly seriously by those involved. Rowers go out on the water whatever the weather, come rain or snow, breaking the ice on freezing winter mornings and sweating in the early summer heat. Rowing requires dedication and commitment, and that is the real reason for its popularity at university. It acts, at least for some students, as a perfect counterpoint to the stresses and demands of study. In Hawking's case it was the perfect remedy for a calcifying boredom with everything else Oxford had to offer.

Rowing is one of the most physically demanding sports around and an oarsman simply has to be powerfully built to help move a boat through the water, but there is one other essential ingredient in every rowing team – the coxswain, or 'cox'.

Hawking was perfectly suited to coxing. He was light so that he did not burden the boat, and he had a loud voice with which he enjoyed barking instructions the length of the boat and enough discipline to attend all the training sessions. His rowing trainer was Norman Dix, who had been with the University College Rowing Club for decades. He recalls that Hawking was a competent enough cox, but never interested in going beyond the second eight. He suspects that the first crew held little appeal because it meant taking it all too seriously, and at that level the fun would have gone out of the whole thing.

Dix remembers Hawking as a boisterous young fellow who from the beginning cultivated a daredevil image when it came to navigating his crew on the river. Many was the time he would return the eight to shore with bits of the boat knocked off and oar blades damaged, because he had tried to guide his crew through an impossibly narrow gap and had come to grief. Dix never did believe Hawking's claims that 'something had got in the way'.

'Half the time,' says Dix, 'I got the distinct impression that he was sitting in the stern of the boat with his head in the stars, working out his mathematical formulae.'

The crews worked hard on the river. They would be out in the boats nearly every day during term time, preparing for the big races, the Torpids, which take place in February, and the Summer Eights in the summer term. The term Torpids originally came from the adjective 'torpid' because this would be the first competitive race in which freshmen could compete, and therefore the standard of many crews was pretty low. Having joined the Rowing Club in October, the novice rowers would have trained hard all winter in preparation for showing off their new-found skills by the fifth week of the winter term.

Torpids are all college 'bumping' races, taking place over several days. The thirteen boats competing start off one hundred and forty feet apart. Each is tied to the bank by a forty-foot line, the other end of which is held by the cox. When the starting-gun goes off, the cox releases the line and the boats chase each other along a stretch of the river with the aim of bumping the boat in front without getting bumped themselves. The main task of the cox is to guide the crew so that they avoid being bumped by the boat behind but manage to bump the boat in front. The object of the exercise is to move up through the positions of the thirteen boats by managing to bump without being bumped; after each heat the 'bumpers' and the 'bumped' change places. If a crew does very well and moves up several places during the series of races, each crew member is entitled to purchase an oar on to which can be written the triumphant tally of bumps, the names of the crew and the date. Such oars adorn the walls of victors' studies. Hawking's crews were pretty average, notching up only a modest number of bumps during their Torpids races, but the whole idea was to have fun and to relieve academic pressures.

After the races came the celebrations and commiserations, both of which would be accompanied by a surfeit of ale, followed by a Rowing Club dinner in the College accompanied by speeches and toasts. And here was the real reason Hawking was involved at all. He was something of a misfit during his first year, lonely and needing to alleviate the boredom of work that presented no challenge to him. The Rowing Club brought the nineteen-year-old out of himself and gave him an opportunity to become part of the university 'in crowd'.

When old schoolfriends encountered Hawking during his second year they could hardly believe the change in him. Variously described as 'one of the lads' and 'definitely raffish', the slender, tousle-haired youth in his pink Rowing Club scarf seemed a far cry from the gawky schoolboy swot who had left St Albans School less than two years earlier. He was no longer a social also-ran, but a fully paid-up member of the 'right' social set. It was very much an all-male domain; women rarely entered this world. It was, in a way, a continuation of the gang at St Albans School, without the intellectual intensity but with a lot more alcohol. The idea was to drink copious amounts of ale, recount vaguely lurid stories and have as much harmless fun as possible. However, his new-found taste for adventure almost got him into trouble.

One night he decided to create a splash. After a few beers with a friend the two of them headed off to one of the foot-bridges spanning the river. After leaving the pub, they picked up a can of paint and some brushes they had left in college and hid them inside a bag. When they arrived at the bridge they set up a couple of wooden planks parallel to the span and suspended them over the water by a carefully arranged rope a few feet below the parapet. Clambering over the side, they positioned themselves on the planks with the can of paint and brushes and began to write. A few minutes later, just visible in the dark, were the words 'VOTE LIBERAL' in foot-high letters along the side of the bridge and clear to anyone on the river when daylight broke.

Then disaster struck. Just as Hawking was finishing off the last letter, the beam of a torch shone down on them from the bridge and an angry voice called out, 'And what do you think you're up to then?' It was a local policeman. The two panicked, and Hawking's friend scurried off the planks and on to the river bank, hightailing it back to town and leaving Hawking with paint brush in hand to face the music. The story goes that he simply got a ticking off from the local constabulary, and the incident was eventually forgotten. But it must have had the desired effect of scaring the life out of him, because he never clashed with the law again.

*

Less than three years after arriving at Oxford University, Hawking again had to face the music when Finals approached and he suddenly found that he could have been better prepared. Dr Berman knew that Hawking, for all his innate ability, would find the examinations harder than he anticipated. Berman realized that there were two types of student who did well at Oxford: those who were bright and worked very hard, and those who had great natural talent and worked very little. It was always the former who achieved greater things in written papers. That was the way of exams; winning end-of-year prizes was one thing, but Finals were in a different league. They were all or nothing, the focal point of the whole three years of study. Hawking once calculated that during the entire three years of his course at Oxford he had done something like 1,000 hours' work, an average of one hour per day – hardly a foundation for the arduous Finals. One friend remembers with amusement, 'Towards the end he was working as much as three hours a day!'

However, Hawking had devised a plan. Because candidates had a wide choice of questions from each paper, he would, he decided, attempt only theoretical physics problems and ignore those requiring detailed factual knowledge. He knew that he could do any theoretical question by using his proven natural talent and intuitive understanding of the subject. But there was an additional problem to complicate things. He had applied to Cambridge to begin PhD studies in cosmology under the most distinguished British astronomer of the day, Fred Hoyle. The catch was that to be accepted for Cambridge he had to achieve a first-class honours degree, the highest possible qualification at Oxford.

The night before Finals Hawking panicked. He tossed and turned all night and got very little sleep. When morning came he dressed up in subfusc, the regulation black gown, white shirt and bow-tie worn by all examinees, left his rooms bleary-eyed and anxious and headed for the examination halls a few yards along The High. Out on the street hundreds of other identically dressed students streamed along the pavements, some clutching books under their arms, others puffing manically on their last cigarette before entering the examination hall – a feast for the tourist's camera but abject misery for those having to sit through days of examinations.

The examination halls themselves do their best to intimidate: high ceilings, great chandeliers hanging down from the void, row upon row of stark wooden desks and hard chairs. Invigilators pace up and down the rows keeping an eagle eye on the candidates in their multitude of poses – staring at the ceiling or the middle distance, pen protruding between clenched teeth, or terminally absorbed, hunched over a manuscript in a free-flow trance. Hawking woke up a little as the paper was placed on the desk before him, and duly followed through his plan of attempting only the theoretical problems.

Exams over, he went off to celebrate the end of Finals with the others of his year, guzzling champagne from the bottle and joining the throng of students ritualistically stopping the traffic on The High and spraying bubbly into the summer sky. After a short pause and a period of nail-biting anticipation, the results were announced. Hawking was on the borderline between a First and a Second. To decide his fate, he would have to face a viva, a personal interview with the examiners.

He was fully aware of his image at the University. He had the impression, rightly or wrongly, that he was considered a difficult student in that he was scruffy and seemingly lazy, more interested in drinking and having fun than working seriously. However, he underestimated how highly thought of were his abilities. Not only that, but as Berman is fond of saying, Hawking was in his element at a viva, because if the examiners had any intelligence they would soon see that he was cleverer than they were. At the interview he made a pronouncement which perfectly encapsulates the man's matter-of-fact attitude and at the same time may have just saved his career. The chief examiner asked him to tell the board of his plans for the future.

'If you award me a First,' he said, 'I will go to Cambridge. If I receive a Second, I shall stay in Oxford, so I expect you will give me a First.'

They did.

4

Doctors and Doctorates

It has been said that Cambridge is the only true university town in England. Oxford is a much larger city and has, lying beyond the ring road, heavy industrial areas nestling next to one of Europe's largest housing estates. Cambridge is altogether quainter and more thoroughly dominated by academia. Although evidence suggests that the University of Cambridge was established by defectors from Oxford, both seats of learning were created at around the same time in the twelfth century, using as their model the University of Paris. Like Oxford, Cambridge University is a collection of colleges under the umbrella of a central university authority. Like Oxford, it attracts the very best scholars from around the world and has a global reputation, paralleled only by its great rival and historical twin a mere eighty miles away. And, like Oxford, it is steeped in tradition, drama and history.

Having just returned from abroad, Stephen Hawking BA (Hon.) arrived in Cambridge in October 1962, exchanging the scorched, barren landscape of the Middle East for autumnal wind and drizzle across the darkening fields of East Anglia. As he travelled past the meadows and gently rolling hills towards his new home that rainy morning, a darkening shadow hung skulking behind the peace and calm of 'the only true university town in England', and indeed behind every other human dwelling elsewhere on the planet, for the world was in the terrifying grip of the Cuba Crisis.

It really did seem as though the world could end in a blaze of nuclear fury at any moment. In these relatively calm post-*glasnost* days, it is perhaps hard to imagine the atmosphere of the time, the insecurity and the uncertainty. Hawking was no different from the next man in experiencing a sense of hopelessness in the face of events over which he had absolutely no control. Old idols, the beautiful and the good, were fading and falling; new heroes stood on the sidelines, ready to emerge. Marilyn Monroe had died in August that year, John F. Kennedy had little more than twelve months to live, and the Beatles were poised on the brink of huge international fame unparalleled in the history of popular culture.

Despite the overbearing threat of imminent annihilation, life in Cambridge went on pretty much as usual. Students began to settle into their new homes and find their feet in a strange city, the townsfolk continued about their daily business as they had done for the thousand years during which the city had existed. In the days leading up to his move to Cambridge, with the world outside looking set to tear itself apart, Stephen Hawking was gradually becoming aware of an inner, personal crisis. Towards the end of his time at Oxford he had begun to find some difficulty in tying his shoelaces, he kept bumping into things, and a number of times he felt his legs give way from under him. Without a drink passing his lips he would, on occasion, find his speech slurring as though he were intoxicated. Not wanting to admit to himself that something was wrong, he said nothing and tried to get on with his life.

Upon arriving in Cambridge another problem arose. When he had applied to do a PhD at the University there were two possible areas of research open to him: elementary particles, the study of the very small, and cosmology, the study of the very large. As he has said himself:

I thought that elementary particles were less attractive, because, although they were finding lots of new particles, there was no proper theory of elementary particles. All they could do was arrange the particles in families, like in botany. In cosmology, on the other hand, there was a well-defined theory – Einstein's general theory of relativity.[1]

However, there was a snag. He had originally chosen to go to Cambridge University because at the time Oxford could not offer cosmological research and, most important, he wanted to study under Fred Hoyle, who had a worldwide reputation as the most eminent scientist in the field. But, instead of getting Hoyle, he was placed under the charge of one Dennis Sciama, of whom he had never heard. For a while this turn of events struck him as disastrous, but he came to realize that Sciama would be a far better supervisor because Hoyle was forever travelling abroad and could find little time to play the role of mentor. He soon discovered too that Dr Sciama was himself a very fine scientist and a tremendously helpful and stimulating supervisor, always available for him to talk to.

Hawking's first term at Cambridge went rather badly. He found that he had not studied mathematics to a sufficiently high standard as an undergraduate and was soon struggling with the complex calculations involved in general relativity. He was still operating in his somewhat lazy work mode, and his research material was becoming increasingly demanding. For the second time in his life he was beginning to flounder. Sciama recalls that, although his student seemed exceptionally bright and ready to argue his point thoroughly and knowledgably, part of Hawking's problem was finding a suitable research problem to study.

The difficulty was that a research assignment had to be sufficiently taxing to meet the requirements of a PhD course, and, because relativity research at that level was fairly new and unusual, the right sort of problem was not easy to find.

Sciama believes that at that time Hawking came close to losing his way and flunking the whole thing. This was a situation which persisted for at least the first year of his PhD. Things would begin to resolve themselves only through a complex series of events initiated by changes already unfolding inside Hawking's own body.

When Stephen returned to St Albans for the Christmas vacation at the end of 1962, the whole of southern England was covered in a thick blanket of snow. In his own mind he must have known that something was wrong. The strange clumsiness he had been experiencing had occurred more frequently but had gone unobserved by anyone in Cambridge. Sciama remembers noticing early in the term that Hawking had a very slight speech impediment but had put it down to nothing more than that. However, when he arrived at his parents' home, because he had been away for a number of months, they instantly noticed that something was wrong. His father's immediate conclusion was that Stephen had contracted some strange bug while in the Middle East the previous summer – a logical conclusion for a doctor of tropical medicine. But they wanted to be sure. They took him to the family doctor who referred him to a specialist.

On New Year's Eve, the Hawkings threw a party at 14 Hillside Road. It was, as might have been expected, a civilized affair with sherry and wine; close friends were invited, including schoolfriends John McClenahan and Michael Church. The word passed around that Stephen was ill, the exact nature of the disease unknown, but something picked up in foreign climes was the general impression. Michael Church remembers that Stephen had difficulties pouring a glass of wine and that most of the liquid ended up on the tablecloth rather than in the glass. Nothing was said, but there was an atmosphere of foreboding that evening.

A young woman called Jane Wilde, whom Stephen had

previously known only vaguely, had also been invited to the party. He was formally introduced to her by a mutual friend during the course of the evening. Jane also lived in St Albans and attended the local high school. As the dying hours of 1962 trickled away and 1963 began, the two of them began to talk and to get to know each other. She was in the upper sixth and had a place at Westfield College in London to begin reading modern languages the following autumn. Jane found the twenty-one-year-old Cambridge postgraduate a fascinating and slightly eccentric character and was immediately attracted to him. She recalls sensing an intellectual arrogance about him, but 'There was something lost, he knew something was happening to him of which he wasn't in control.'[2] From that night their friendship blossomed.

He was due back in Cambridge to begin the Lent term later in January, but instead of resuming his work there he was taken into hospital to undergo a series of investigatory tests. Hawking recalls the experience vividly:

> They took a muscle sample from my arm, stuck electrodes into me, and injected some radio-opaque fluid into my spine, and watched it going up and down with X-rays, as they tilted the bed. After all that, they didn't tell me what I had, except that it was not multiple sclerosis, and that I was an atypical case. I gathered, however, that they expected it to continue to get worse, and that there was nothing they could do except give me vitamins. I could see that they didn't expect them to have much effect. I didn't feel like asking for more details, because they were obviously bad.[3]

The doctors advised him to return to Cambridge and his cosmological research, but that, of course, was easier said than done. Work was not going well and now the ever-present possibility of imminent death hung over his every thought and action. He returned to Cambridge and awaited the results of the tests. A short time later he was diagnosed as having a rare and incurable disease called amyotrophic lateral sclerosis, or ALS, known in the United States as Lou Gehrig's Disease after the Yankee baseball player who died from the illness. In Britain it is usually called motor neuron disease.

ALS affects the nerves of the spinal cord and the parts of the brain which produce voluntary motor functions. The cells gradually degenerate over a period of time and cause paralysis as muscles atrophy throughout the body. Apart from this the brain is unaffected, and the higher functions such as thought and memory are left untouched. The body gradually wastes away but the patient's mind remains intact. The usual prognosis is gradual immobility, followed by creeping paralysis leading eventually to death by suffocation or pneumonia as the respiratory muscles seize up. The symptoms are painless, but in the final stages of the disease patients are often given morphine to alleviate chronic depression.

One of the amazing ironies of the situation was that Stephen Hawking just happened to be studying theoretical physics, one of the very few jobs for which his mind was the only real tool he needed. If he had been an experimental physicist, his career would have been over. Quite naturally this was little compensation to the twenty-one-year-old who, like everyone else, had seen a normal life stretching ahead of him rather than a death sentence from a neurological disease. The doctors had given him two years.

Upon hearing the news, Hawking fell into a deep depression. Fleet Street legend has it that he locked himself away in a darkened room, plummeting into heavy drinking and listening to a great deal of high-volume Wagner while wallowing in a drunken haze of self-pity. However, he has gone on record as saying that the stories of excessive drinking are exaggerated, but that, feeling a somewhat 'tragic character',[4] he did shut himself away for a while and listened to a lot of music, especially Wagner.

Reports in magazine articles that I drank heavily are an exaggeration. The trouble is, once one article said it, other articles copied it, because it made a good story. Anything that has appeared in print so many times must be true.[5]

The truth may never be known, but Hawking's recollection of events rings true. The idea of getting totally smashed and staying that way to nullify the mental pain strikes one as an eminently reasonable thing to do in the circumstances. Furthermore, there is evidence to support his assertion. Dennis Sciama, for one, has said that he has no recollection of Hawking disappearing for a long period, as the tabloids have implied. Being used to seeing his students every day during term time, he would have been the first to have noticed Hawking's absence.

However, there is little doubt that he was deeply shocked by the news and experienced a time of deep depression. There seemed very little point in continuing with his research because he might not live long enough to finish his PhD. For a while he quite naturally believed that there was nothing to live for. If he was going to die within a few years, then why bother to do anything now? He had never been attracted by religion or any thought of an afterlife, so there was no crumb of comfort to be found there. He would live his span and then die. That was his fate. Being no different from the next person faced with any form of personal tragedy, he kept thinking, 'How could something like that happen to me? Why should I be cut off like this?'[6]

He tells of an experience while he was undergoing tests that made a great impression on him and helped him through those nightmare days back in Cambridge:

While I had been in hospital, I had seen a boy I vaguely knew die of leukemia, in the bed opposite me. It had not been a pretty sight. Clearly there were people who were worse off than me. At least my condition didn't make me feel sick. Whenever I feel inclined to be sorry for myself, I remember that boy.[7]

He was experiencing some disturbing but poignant dreams at the time. In hospital he dreamt that he was going to be executed. He suddenly realized that there were a lot of worthwhile things he could do if he were to be reprieved. In another recurring dream he thought that he could sacrifice his life to save others: 'After all, if I were going to die anyway, it might as well do some good,' he dreamt.[8]

After Hawking had dragged himself out of his depression and back to work, his father decided to pay Dennis Sciama a visit. He explained the situation and asked if Stephen could complete his PhD in a shorter time than the three-year minimum because his son might not live that long. Sciama, knowing perhaps better than anyone what his student was really capable of, told Frank Hawking that any idea of finishing in less than three years was impossible. Whether Sciama realized at the time that Hawking would need his work to help him through is another matter; but he knew the rules, and despite the fact that his student may have been dying, they could not be bent to suit him.

Most people believed that the medical predictions were correct and that Hawking had a very short time to live. John McClenahan vividly remembers that, on the eve of his departure to work in America for a year, Hawking's sister Mary had said to him that, if he decided not to return within a year, he would probably never see his friend again. Once it had taken a grip, the disease developed quickly. Jane met Stephen again soon after he was released from hospital, and found him confused and lacking the will to live.

However, there is little doubt that her appearance on the scene was a major turning point in Hawking's life. The two of them began to see a lot more of one another, and a strong relationship developed. It was finding Jane that enabled him to break out of his depression and regenerate some belief in his life and work. Meanwhile, the PhD progressed at a painfully slow rate.

He was not the only student working with Sciama. A South African, George Ellis, had been the supervisor's first student when Sciama had taken up his post in 1961. A year later Hawking arrived, followed the year after by two other students who would, along with Ellis, become lifelong friends and colleagues – Brandon Carter and Martin Rees. Together with a number of others they formed a small group of relativists and cosmologists, all working on slightly different areas within the same field.

They became good friends as well as co-workers, often relaxing in one of the city's pubs in the evening or going to concerts, plays and films together when they had had enough of talking physics over a pint of beer. There were common interests other than their work. Ellis was always very interested in politics and vehemently anti-apartheid. In Hawking he found a sympathetic set of attitudes, and they would often talk politics. Sitting beside pub fires in the winter, and in gardens on summer evenings, the two of them would discuss anything, from the Vietnam War to Black Power. They were all introduced to Jane, of course, and when she made the trip to Cambridge at weekends the whole group would often go out together to eat or to picnic by the river, watching the punts glide by.

During Hawking's first year he worked with the other students and supervisors in the Phoenix Wing of the Cavendish Laboratory, which had been set up by James Clerk Maxwell in the 1870s. In the early 1960s the Head of the Physics Department, George Batchelor, managed to persuade the University to establish a separate mathematics and theoretical physics department in what used to be known as the Old University Press Building in Silver Street. It became known as the Department of Applied Mathematics and Theoretical Physics (DAMTP).

The system at Cambridge is such that both undergraduate and postgraduate students are enrolled in one of the colleges, yet work in university buildings with others in the same field but from different colleges. Hawking was a student of Trinity Hall and would eat there in the evenings and be assigned accommodation by the college, but he did not work in Trinity Hall buildings or exclusively with Trinity Hall students and academic staff.

The atmosphere in the Physics Department was very informal, and PhD students had no rigid timetable or course to follow. The job of the supervisor is to suggest a set of problems or targets, and discuss with the student plans of attack and give guidance where necessary. Sciama remembers how, on a number of occasions, he would dash into Hawking's office with a new idea for something his charge was working on, and they would then thrash out the scheme together. At other times Hawking would go to see Sciama in his office, a fondly remembered place, the walls covered with modern art prints between the shelves of books and papers.

As well as attending lectures at the University, all the PhD students at the DAMTP attended regular seminars, where thirty or forty people would listen to talks given by one of the teaching staff or a visiting lecturer. These would be followed by a general discussion. But the most important place for conversation and exchanging ideas was in the Tea Room. In the twice-daily ritual, well established at the Cavendish and carried over to Silver Street, everyone would meet at 11 a.m. for coffee and 4 p.m. for tea to exchange their latest thoughts and ideas. Students shared offices, and their doors were nearly always open to all – there was never any feeling of working secretly or keeping ideas to oneself. It was in this atmosphere of free communication that Hawking happened to stumble upon his first significant project during his early years as a PhD student.

Fred Hoyle was a very big name in the Physics Department of Cambridge University, widely known for his ideas about the origins of the Universe. An inveterate self-publicist, he was very good at manipulating the media and was of the breed of scientist who would on occasion publicly express unrefereed and unverified theories. His justification for this was simple. He was not an ego-maniac or intellectual cowboy, but to acquire funds for his research he needed to make a public splash, to be internationally famous. Publicity was of the utmost importance to him.

Hoyle had not always been in such an elevated position. The son of a Yorkshire textile merchant, he had entered Cambridge in the 1930s on a full scholarship and had been hardened by the experience of feeling socially inadequate because of his background and strange accent. Although he proved himself intellectually superior to most of his contemporaries, he was changed by the experience and emerged as a difficult customer to deal with. For much of his time as a professor at Cambridge he was engaged in fierce arguments with the authorities as well as many of his colleagues. Soon after the move to Silver Street, Hoyle set up his own institute in Cambridge, but still used the brains and help of many at the DAMTP.

During the arguments and upheavals at Cambridge Hoyle was very much involved with the steady state theory of the Universe. He had developed the idea with the mathematician Hermann Bondi at King's College, London, and the astronomer Thomas Gold, but at the time it was simply the more scientifically evolved of two contending theories. He detested the alternative theory of a spontaneous creation of the Universe, which he once described as a party girl jumping out of a birthday cake – it just wasn't dignified or elegant. Much to his later amusement, he became the creator of the term 'Big Bang', a phrase coined deliberately to ridicule the idea and dropped into a radio programme in which he was propounding his own steady state theory.

As well as developing his theory of the origin of the Universe, Hoyle also acted as supervisor to a select group of students. One of his charges was a graduate student called Jayant Narlikar. Narlikar had been assigned the task of working through some of the mathematics for Hoyle's theory as part of the research material for his PhD. He also happened to occupy the office next to Hawking's. Hawking became very interested in Narlikar's equations. Without too much persuasion, Narlikar shared the research material he was working on and Hawking began to develop the theories further. During the next few months Hawking spent more and more time walking between his friend's office and his own, clutching pages full of mathematical interpretations in one hand and leaning heavily on his newly acquired walking stick with the other.

At this point it should be emphasized that Hawking had no malicious intent towards Hoyle or, indeed, Narlikar. He was quite simply curious about the material and was floundering with his own projects. The equations and their meaning were fascinating, and perhaps initially more stimulating than his own research. Besides which, the whole approach within the department was one of shared goals and ideals.

Before too long things came to a head. Hoyle decided to make a public announcement of his findings at a meeting of The Royal Society in London. Although it was certainly not without precedent, some of his colleagues considered that he was being over-keen in doing this because the work had not been refereed. Hoyle gave his talk to around a hundred people; at the end there was warm applause and the usual post-lecture hubbub of conversation. Then he asked if there were any questions. Naturally Hawking had attended and had followed the arguments closely. He stood up slowly, clutching his stick. The room fell silent.

'The quantity you're talking about diverges,' he said.

Subdued murmurs passed around the audience. The gathered scientists saw immediately that if Hawking's assertion were correct, then Hoyle's latest offering would be shown to be false.

'Of course it doesn't diverge,' Hoyle replied.

'It does,' came Hawking's defiant reply.

Hoyle paused and surveyed the room for a moment. The audience was absolutely silent. 'How do you know?' he snapped.

'Because I worked it out,' Hawking said slowly.

An embarrassed laugh passed through the room. This was the last thing Hoyle wanted to hear. He was furious with the young upstart. But any enmity between the two men was short-lived – Hawking had demonstrated himself to be too good a physicist for that. But Hoyle considered Hawking's action to be unethical and told him so. In return, Hawking and others pointed out that Hoyle had been unethical in announcing results which had not been verified. The only innocent party, who no doubt had to bear the full brunt of Hoyle's anger, was the middle-man, Narlikar.

Although Hoyle is every bit Hawking's intellectual equal, on this occasion the younger man turned out to be absolutely correct: the quantity Hoyle had been talking about did indeed diverge, which meant that the latest component of his theory was wrong. Hawking wrote a paper summarizing the mathematical findings which had led him to realize this. It was well received by his peers and established him as a promising young researcher. While still trying to sort out his own PhD work with Sciama, Hawking was already beginning to make a name for himself within the rarefied atmosphere of cosmological research.

During his first two years at Cambridge, the effects of the ALS disease rapidly worsened. He was beginning to experience enormous difficulty in walking, and was compelled to use a stick in order to move just a few feet. His friends helped him as best they could, but most of the time he shunned any assistance. Using walls and objects as well as sticks, he would manage, painfully slowly, to traverse rooms and open areas. There were many occasions when these supports were not enough. Sciama and his colleagues remember clearly that on some days Hawking would turn up at the office with a bandage around his head, having fallen heavily and received a nasty bump.

His speech was also becoming seriously affected by the disease. Instead of being merely slurred, his speaking voice was now rapidly becoming unintelligible, and even close colleagues were experiencing some difficulty in understanding what he was saying. Nothing slowed him down, however; in fact, he was just getting into his stride. Work was progressing faster and more positively than it had ever done in his entire career, and this serves to illustrate his attitude to his illness. Crazy as it may seem, ALS is simply not that important to him. Of course he has had to suffer the humiliations and obstructions facing all those in our society who are not able-bodied, and naturally he has had to adapt to his condition and to live under exceptional circumstances. But the disease has not touched the essence of his being, his mind, and so has not affected his work.

More than anyone else, Hawking himself would wish to underplay his disability and to concentrate on his scientific achievements, for that is really what is important to him. Those working with him, and the many physicists around the world who hold him in the highest regard, do not view Hawking as anything other than one of them. The fact that he cannot now speak and is immobile without the technology at his fingertips is quite irrelevant. To them he is friend, colleague and, above all, great scientist.

Having come to terms with ALS and found someone in Jane Wilde with whom he could share his life on a purely personal level, he began to blossom. The couple became engaged, and the frequency of weekend visits increased. It was obvious to everyone that the two of them were sublimely happy and immensely important to each other. Jane recalls, 'I wanted to find some purpose to my existence, and I suppose I found it in the idea of looking after him. But we were in love.'[9] On another occasion she said, 'I decided what I was going to do, so I did. He was very, very determined, very ambitious. Much the same as now. He already had the beginnings of the condition when I first knew him, so I've never known a fit, able-bodied Stephen.'[10]

For Hawking his engagement to Jane was probably the most important thing that had ever happened to him: it changed his life, gave him something to live for and made him determined to live. Without the help that Jane gave him, he almost certainly would not have been able to carry on, or had the will to do so.

From this point on his work went from strength to strength, and Sciama began to believe that Hawking might, after all, manage to bring together the disparate strands of his PhD research. It was still touch and go, but another chance encounter was just around the corner.

Sciama's research group became very interested in the work of a young applied mathematician, Roger Penrose, who was then based at Birkbeck College in London. The son of an eminent geneticist, Penrose had studied at University College in London and had gone on to Cambridge in the early fifties. After research in the United States he had begun in the early sixties to develop ideas of singularity theory which interfaced perfectly with the ideas then emerging from the DAMTP.

The group from Cambridge began to attend talks at King's College in London where the great mathematician and co-creator of the steady state theory, Hermann Bondi, was Professor of Applied Mathematics. King's acted as a suitable meeting-point for Penrose (who travelled across London), those from Cambridge, and a small group of physicists and mathematicians from the college itself. Sciama took Carter, Ellis, Rees and Hawking to the meetings with the idea that the discussions might spark off applications to their own work. However, there were times when Hawking almost failed to make it to London.

Brandon Carter remembers one particular occasion when the group arrived late at the railway station and the train was already drawing in. They all ran for it, forgetting about Stephen, who was struggling along with his sticks. It was only after they had installed themselves in the carriage that they were aware he was not with them. Carter recalls looking out of the window, seeing a pathetic figure struggling towards them along the platform and realizing that Stephen might not make it before the train pulled away. Knowing how Hawking was fiercely against being treated differently from others, they did not like to help him too much. However, on this occasion Carter and one of the others jumped out to help him along the platform and on to the train.

It would have been an odd twist of fate indeed if Hawking had not made it to at least one of those London meetings because it was through them that his whole career took another positive turn. Over the course of the talks at King's, Roger Penrose had introduced his colleagues to the idea of a spacetime singularity at the centre of a black hole, and naturally the group from Cambridge were tremendously excited by this.

One night, on the way back to Cambridge, they were all seated together in a second-class compartment and had begun to discuss what had been said at the meeting that evening. Feeling disinclined to talk for a moment, Hawking peered through the window, watching the darkened fields stream past and the juxtaposition of his friends reflected in the glass. His colleagues were arguing over one of the finer mathematical points in Penrose's discussion. Suddenly, an idea struck him, and he looked away from the window. Turning to Sciama sitting across from him, he said, 'I wonder what would happen if you applied Roger's singularity theory to the entire Universe.' In the event it was that single idea that saved Hawking's PhD and set him on the road to science superstardom.

Penrose published his ideas in January 1965, by which time Hawking was already setting to work on the flash of inspiration that had struck him on the way home from London to Cambridge that night after the talk. Applying singularity theory to the Universe was by no means an easy problem, and within months Sciama was beginning to realize that his young PhD student was doing something truly exceptional. For Hawking, this was the first time he had really applied himself to anything. As he says:

> I . . . started working hard for the first time in my life. To my surprise, I found I liked it. Maybe it is not really fair to call it work. Someone once said, 'Scientists and prostitutes get paid for doing what they enjoy.'[11]

When he was satisfied with the mathematics behind the ideas, he began to write up his thesis. In many respects it ended up as a pretty messy effort because he had been in something of a wilderness for much of the first half of his time at Cambridge. The problems he and Sciama had experienced in finding him suitable research projects left a number of holes and unanswered questions in the thesis. However, it had one saving grace – his application of singularity theory during his third year.

The final chapter of Hawking's thesis was a brilliant piece of work and made all the difference to the awarding of the PhD. The work was judged by an internal examiner, Dennis Sciama, and an expert external referee. As well as being passed or failed, a PhD can be deferred, which means that the student has to resubmit the thesis at a later date, usually after another year. Thanks to his final chapter, Hawking was saved this humiliation and the examiners awarded him the degree. From then on the twenty-three-year-old physicist could call himself Dr Stephen Hawking.

5

From Black Holes to the Big Bang

In the early 1960s, astronomers already knew that any star which contains more than about three times as much matter as our Sun ought to end its life by collapsing inward to form what is now known as a black hole. More than twenty years previously, researchers had used Einstein's equations of general relativity to calculate that such an object would bend spacetime completely round upon itself, cutting the central mass off from the rest of the Universe. Light rays passing near such an object would be deflected so much that even photons would orbit around the central 'star' in closed loops, and could never escape into the Universe outside. Obviously, since it could emit no light, such an object would be black, which is why the American relativist John Wheeler dubbed them 'black holes' in 1969.

But although it was well known that the general theory made this prediction, at the time Hawking was completing his undergraduate studies and moving on to research, no one took the notion of black holes seriously. The reason is that there are very many known stars which have more than three times the mass of our Sun. They do not collapse, because nuclear reactions going on inside the stars make them hot. The heat creates an outward pressure which holds the star up against the pull of gravity. Astronomers knew that when such stars run out of nuclear 'fuel', they explode, blasting away their outer layers into space. As recently as thirty years ago, astronomers assumed that such an explosion would always blow away so much matter that the core left behind would have less than three times the mass of our Sun – or, perhaps, that some as yet undiscovered pressure would come into play as the remnant of star-stuff began to shrink.

This prejudice was reinforced by the fact that astronomers had indeed discovered many old, dead stars. These stellar cinders all had a bit less than the mass of our Sun, but compressed into a volume only about as big as that of the Earth. Such planet-sized stars are known as white dwarfs. They are held up against the inward pull of gravity by the pressure of the electrons associated with the atoms of which they are made, acting like a kind of electron gas. A white dwarf is so dense that each cubic centimetre of the star contains a million grams of material. Before 1967, these were the densest known objects in the Universe.

But although astronomers did not seriously believe that anything denser than a white dwarf could exist, a few mathematicians enjoyed playing with Einstein's equations to work out what would happen to matter if it were squeezed to still greater densities. The equations said that if three times as much matter as our Sun contains were squeezed until it occupied a spherical region with a radius of just under 9 kilometres, spacetime in its vicinity would be so distorted that not even light could escape. Because nothing can travel faster than light, this meant that nothing at all could ever escape from such an object, which the mathematicians sometimes referred to as a 'collapsar' (from collapsed star). It would have become the ultimate bottomless pit, into which anything could fall but from which nothing could ever emerge. And the density inside the collapsar would be greater than the density of the nucleus of an atom; this, theorists of the time thought, was clearly impossible.

In fact, they did consider (but not too seriously) the possibility of stars as dense as the nucleus of an atom. By the 1930s, physicists knew that the nucleus of an atom is made of closely packed particles called protons and neutrons. The protons each carry one unit of positive charge; the neutrons, as their name suggests, are electrically neutral, but each has about the same mass as a proton. In everyday atoms, like the ones this book is made of, each nucleus is surrounded by a cloud of electrons. Each electron carries one unit of negative charge, and there is the same number of electrons as protons, so the atom as a whole is electrically neutral.

But an atom is largely empty space. The nucleus is tiny, but very dense, and the cloud of electrons is (by comparison) huge and insubstantial. In proportion to the size of a whole atom, the nucleus is like a grain of sand in the middle of a concert hall. In white dwarf stars, some of the electrons are knocked off their atoms by the high prevailing pressure, and the nuclei are embedded in a sea of electrons which belongs to the whole star, not to any particular nucleus. But there is still a lot of space between the nuclei, even though that space contains electrons. Each nucleus has positive charge, and like charges repel, so the nuclei keep their distance from each other.

But quantum theory said that there is a way to make a star more dense than a white dwarf. If the star were squeezed even more by gravity, the electrons could be forced to combine with protons to make more neutrons. The result would be a star made entirely of neutrons, and these could be packed together as closely as the protons and neutrons in an atomic nucleus. This would be a neutron star.

Calculations suggested that this ought to happen for any dead star with a mass more than 20 per cent larger than that of our Sun (that is, more than 1.2 solar masses). A neutron star would have that much mass packed within a radius of about 10 kilometres, no bigger than many mountains on Earth. The density of the matter in a neutron star, in grams per cubic centimetre, would be 10^{14} – that is, 1 followed by 14 zeros, or one hundred thousand billion. Even an object this dense would not be a black hole, though, for light could still escape from its surface into the Universe at large.

Making a black hole from a dead star would require, as the theorists of the early 1960s were well aware, crushing even neutrons out of existence. The quantum equations said, in fact, that there was no way that even neutrons could hold up the weight of a dead star of 3 solar masses or more, and that if any such object were left over from the explosive death throes of a massive star, it would collapse inwards completely, shrinking to a mathematical point called a singularity. Long before the collapsing star could reach this state of zero volume and infinite density, it would have wrapped spacetime around itself, cutting off the collapsar from the outside Universe.

Indeed, the equations said that if you squeezed *any* collection of matter hard enough it would collapse in this way. The special feature of objects of more than 3 solar masses is that they will collapse anyway, under their own weight. But if it were possible to squeeze our own Sun down into a sphere with a radius of about 3 kilometres, it would become a black hole. So would the Earth, if it were squeezed down to about a centimetre. In each case, once the object had been squeezed down to the critical size, gravity would take over, closing spacetime around the object while it continued to shrink away into the infinite density singularity inside the black hole. But notice that it is much easier to make a black hole if you have a lot of mass. The critical size is not simply proportional to the amount of mass you have; the density at which a black hole forms is larger if you have less mass to squeeze.

For any mass there is a critical radius, called the Schwarzschild radius, at which this will occur. As these examples indicate, the Schwarzschild radius is smaller for less massive objects – you have to squeeze the Earth harder than the Sun, and the Sun harder than a more massive star, in order to make a black hole. Once it had formed, there would be a surface around the hole (a bit like the surface of the sea) marking the boundary between the Universe at large and the region of highly distorted spacetime from which nothing could escape. It would be a one-way horizon (unlike the surface of the sea!) across which both radiation and material particles could happily travel inward, tugged by gravity to join the accumulating mass of the singularity, but across which nothing at all, not even light, could travel outwards.

Some mathematicians worried, thirty years ago, about the prediction that black holes must contain singularities. The notion of a point of infinite density made them uneasy. But most astronomers were more pragmatic. First of all, they doubted whether black holes could really exist at all. Probably, they thought, some law of physics would prevent any dead star from having enough leftover mass to collapse in this way. And even if black holes did exist, by their very nature they would keep the singularities at their hearts locked away from sight or investigation. Did it really matter, after all, if theory said that points of infinite density could exist, if the same theory said that such singularities were safely locked away behind uncrossable horizons?

One thing, however, should have worried those astronomers, even in the early 1960s. Just as you need to squeeze a small mass hard to make a black hole, a larger mass needs less of a squeeze to do the same trick. Indeed, a mass of about 4.5 billion solar masses would form a black hole if it were all contained within a sphere only twice the diameter of our Solar System. That mass sounds ludicrous, at first. But remember that there are a hundred billion stars in our Milky Way Galaxy. If just 5 per cent of the total mass were involved, such a supermassive black hole could indeed form. And the density of such an object would be nothing like the density of the nucleus of an atom, or a neutron star. It would be just 1 gram per cubic centimetre – the same density as water. You could actually make a black hole out of water, if you had enough of it!

One way to understand how this can happen is by an analogy with running tracks. The important thing about a black hole is that it bends spacetime completely around itself, so that light rays at the horizon would circle endlessly around the central singularity. But the photon 'orbits' can be either very tight or follow a gentle curve. Indoor running tracks are usually tightly curved, to make them fit into the space available. Outdoor running tracks are more gently curved, and take up more space. But in both cases, if you run round the track you get back to where you started from – you follow a closed loop. Similarly, a black hole can be very small, with spacetime tightly folded around itself, or very large, with light rays following gradual curves around the horizon (or, indeed, they can be any size in between).

Very slowly, during the 1960s, the implications of this began to dawn on cosmologists. The whole Universe, they realized, might behave in some ways like the biggest black hole of them all, with everything in the Universe held together by gravity, and all of spacetime forming a self-contained, closed entity that folded round on itself with the ultimate in gradual curvature. But there is one big difference – black holes pull matter inwards, towards the singularity; the Universe expands, outwards from the Big Bang. The Universe is like a black hole inside out.

Einstein's equations – the general theory of relativity – said that the Universe could not be static, but must be either expanding or contracting. Observations showed that the Universe is, indeed, expanding. So what did Einstein's equations say about conditions long ago, when galaxies were packed tightly together, and before? Taken at face value, the equations said that the Universe must have emerged from a point of infinite density, a singularity, about 15 billion years ago. 'Obviously' (to astronomers of the 1940s and 1950s, that is) that was ridiculous. The fact that the equations predicted a singularity must mean that they were flawed in some way; no doubt in due course somebody would come up with a better theory, one that didn't make such extreme predictions. But meanwhile it seemed fairly reasonable to take the equations at face value as far as they applied to conditions that bore some resemblance to those we observe today.

The densest form of matter familiar to us today is nuclear matter: protons and neutrons packed together in the hearts of atoms. So a few brave souls were prepared to contemplate the possibility that the general theory might provide a good guide to how the Universe had evolved from a state in which the overall density was as great as that of the nucleus of an atom, a 'primeval atom', if you like, containing *all* the mass of the Universe in a kind of neutron superstar.

What came 'before' that? How did this primeval superdensity – sometimes referred to as the 'cosmic egg' – come into being? Nobody knew; they could only make guesses. Perhaps the cosmic egg had existed for all eternity, until something triggered it into expansion. Or perhaps there had been a previous phase of the Universe in which spacetime was collapsing, in line with Einstein's equations. Such a contracting universe might compress itself to nuclear densities and then 'bounce' outwards again, into a phase of expansion, without encountering the troublesome singularity.

The notion of the primeval atom, or cosmic egg, emerged in the early 1930s, and was refined over the next couple of decades. Even at the beginning of the 1960s, however, this was all still just a mathematical game played by a few experts, as much as anything for their own amusement. The notion of a superdense cosmic egg, only about thirty times bigger than our Sun but containing everything, which had burst asunder to create the expanding Universe, fitted Einstein's equations and the observations. But nobody seems to have felt, deep down in their hearts, that their equations described *the* Universe. Nobody would have been too worried if it had turned out that the whole idea of the cosmic egg was wrong.

You can get a feel for the way people regarded the idea in the 1950s from their own shorthand terms for describing their work. The equations of the general theory of relativity actually allow for more than one possible description of the overall behaviour of spacetime. As we have mentioned, either expansion or contraction (but not stasis) is allowed by the equations. Obviously, the Universe we live in cannot be expanding and contracting at the same time; the two solutions to the equations cannot both apply to the Universe today. So the solutions are called models. A cosmological model is a set of equations that describes how a universe (with a small 'u') might behave. The equations have to obey the known laws of physics, but they do not necessarily purport to describe the actual behaviour of the real Universe (with a capital 'U'). Both the expanding and the contracting solutions to Einstein's equations describe model universes, intriguing mathematical toys; the expanding solution might describe the real Universe. At the beginning of the 1960s, however, most cosmologists would have preferred to call even the expanding solution simply a model universe.

But during the 1960s the whole notion of the Big Bang, as the theory was known, firmed up. Cosmologists began to believe, as more evidence came in confirming the accuracy of the predictions implicit in the general theory of relativity, that their equations really did describe what was going on out there in the real Universe. This encouraged more theoretical calculations, leading to new predictions, and more observations, in a self-stimulating upward spiral that led to a dramatic revolution in our understanding of the birth of the Universe. By 1976, the Big Bang theory was so well established that American physicist Steven Weinberg was able to write a best-selling popular book, *The First Three Minutes*, describing the early stages of the Big Bang, how the Universe had emerged from the superdense state of the cosmic egg. Although written in the 1970s, the book encapsulated what was essentially the 1960s understanding of the Big Bang; we will not be getting too far ahead of our story if we give a brief résumé of that understanding now.

One of the strangest things to grasp about all these descriptions of the Universe – the relativistic cosmological models – is that the Big Bang does not consist of a huge primeval atom sitting in empty space and then exploding outwards. Many people still have this image, in which the galaxies are like fragments of an exploding bomb, hurtling outwards through space. But it is wrong.

What Einstein's equations tell us is that it is space itself that expands, taking galaxies along for the ride. Galaxies were closer together long ago, when the Universe was younger, because the distances between them were more compressed than they are today. You can see this by imagining two spots of paint on a strip of elastic, or on a rubber band. When you pull on the ends of the strip, it stretches, and the two paint spots move apart, but they do not move through the material the strip is made of.

So, in the very early Universe, at the time of the explosion of the primeval atom, there was no 'outside' for the fragments of the explosion to move into. Space was tightly wrapped around itself, so that the cosmic egg was a completely self-contained ball of matter, energy, space and time. It was, indeed, a superdense black hole. It still is a black hole – the only difference is that, by expanding, it has become a very low density black hole, in which light now follows very gently curving orbits at the horizon.

We live inside a black hole, but one so huge that the bending of spacetime within it is too small to be measured by any astronomical instruments on Earth. The 'explosion' of the Big Bang stretched space, literally creating more room in which the material components of the cosmic egg could move. Starting out very hot and dense, the fireball thinned and cooled as the space available expanded. The process is exactly the same as the way the fluid in the pipes of your refrigerator keeps the fridge cool. In the fridge, fluid expands into a large chamber, and cools; at the back of the fridge, it is squeezed into a smaller space and gets hot, but the heat escapes from the piping on the outside before the fluid goes back into the fridge to repeat the cycle. Like that fluid being squeezed, or like air being compressed in a bicycle pump when we use it to inflate a tyre, the Universe must have been much hotter when it was more compressed.

How much hotter? If you run your cosmological model all the way back to the singularity predicted by Einstein's equations, you would be dealing with infinite temperatures, as well as infinite density. But nobody in the 1960s went to that extreme. The infinities were still taken as indicating a breakdown in the general theory of relativity, but even so the moment at which the infinities occurred in the models could be used as a marker for the beginning of time (at least until someone came up with a better theory).

Although the physics of the 1960s could not say what went on during the split second following that beginning of time, it could describe in great detail everything that had happened to the Universe in the 15 billion years beginning just one-tenth of a second later. To an increasing number of cosmologists, the general theory did not really seem such a bad description of the Universe, if it could explain everything that has happened in the past 15 billion years except for the very first one-tenth of a second. This is what it told them.

One-tenth of a second after 'the beginning' (or after the 'bounce', as many cosmologists of the 1960s would have argued), the density of the Universe was 30 million times greater than the density of water. The temperature was 30 billion degrees,* and the Universe consisted of a mixture of very high energy radiation (photons) and material particles including neutrons, protons and electrons, but also more exotic, unstable particles created ephemerally out of pure radiation. This is the ultimate example of the equivalence between mass and energy, expressed in Einstein's famous equation $E = mc^2$. On the Earth, in an atomic bomb, and inside the Sun, where nuclear reactions take place, tiny amounts of matter (m) are converted into large amounts of energy (E), because c is the speed of light, which is 300,000 kilometres a second, and c^2 is a very large number indeed. But if you had enough energy to play with, you could actually make matter out of energy; and there was ample energy available to do the trick in the Big Bang – even if many of the particles created in this way were unstable, destined to disappear again in a puff of radiation in far less than the blink of an eye.

One second later, 1.1 seconds after the beginning, the Universe had cooled dramatically – all the way down to ten billion K. At that time, the density was just 380,000 times the density of water, and from then on the reactions between particles were very similar to the nuclear reactions that go on inside the Sun and other stars today.

At a temperature of three billion K, just under 14 seconds from the beginning, the first nuclei of deuterium could form, albeit temporarily. Hydrogen is the simplest atom, with just a single proton in its nucleus and one electron orbiting outside the nucleus. (In a sense, lone protons can be regarded as

* Physicists measure temperature in degrees kelvin, denoted by the letter K. This scale of measurement starts from the absolute zero of temperature, at −273°C, where all thermal motion of atoms stops. But a little matter of 273 degrees is neither here nor there when we are measuring temperatures in billions of degrees, so for all practical purposes the temperatures given for the fireball are the same as degrees Celsius.

nuclei of hydrogen atoms.) The next most complicated atom is deuterium, which has a nucleus composed of one proton and one neutron, still with a single electron orbiting around it. Atoms that have the same number of electrons as each other but different numbers of neutrons still have identical chemical properties, and are known as isotopes; deuterium is an isotope of hydrogen, and is often known as 'heavy hydrogen'.

Temperature is a measure of how fast, on average, the particles that make up matter are moving (which is why there can be no temperature colder than −273°C, at which atomic motion stops), and at temperatures above three billion K protons and neutrons move too fast to do anything except bounce off each other. Some particles move faster than the average for a particular temperature, and some slower, although most have speeds close to the average. So, as the temperature fell below that value, some protons and neutrons were moving slowly enough to stick together when they collided. The thing that makes them stick is an attraction known as the strong force. As its name suggests, this is a powerful force of attraction that operates between all protons and neutrons. But it has a very short range, and fast-moving particles brush past or bounce off each other too quickly for the strong force to take hold of them during the brief time they are in range. At first, most of the deuterium nuclei produced in this way were knocked apart by collisions with faster-moving particles, but as the fireball cooled still further the deuterium nuclei had a better and better chance of survival.

Just 3 minutes and 2 seconds after the beginning, the temperature had cooled to below one billion K – the entire Universe was then only seventy times as hot as the heart of the Sun is today. At this point, almost all the deuterium nuclei were able to combine in pairs to form nuclei of helium. These helium nuclei each contain two protons and two neutrons, making four 'nucleons' in all, so they are known as helium-4 nuclei (and helium atoms, of course, each have two electrons orbiting around the nucleus).

It happens that helium-4 nuclei are particularly stable. But there are no stable nuclei containing five nucleons (such as you might expect to get if you added a proton or a neutron to a nucleus of helium-4) or eight nucleons (such as you might expect to get if you stuck two helium-4 nuclei together). So the process of 'nucleosynthesis' in the Big Bang stopped with the production of helium-4. Less than 4 minutes after the beginning, matter had settled down into a mixture of about 75 per cent hydrogen nuclei and 25 per cent helium, intermingling with fast-moving electrons and bathed in a sea of hot radiation.

Half an hour later, 34 minutes after the beginning, the temperature was down to 300 million K, and the density of the Universe was only 10 per cent of the density of water. But it took a further 700,000 years for the Universe to cool enough to allow electrons to become attached to the nuclei and form stable atoms. Before then, as soon as a positively charged nucleus tried to latch on to a negatively charged electron, the electron would have been knocked away by an energetic photon. After 700,000 years, however, the temperature of the Universe had fallen to about 4,000 K (roughly the temperature at the surface of the Sun today), and nuclei and electrons were at last able to hold together to form stable atoms.

For most of the past 15 billion years, protons, neutrons and electrons have been bound up in stars and galaxies formed out of this primeval stuff as gravity pulled clouds of gas together in space. The radiation left over from the Big Bang had nothing more to do with the matter, once it was no longer hot enough to separate electrons from their atomic nuclei, and simply cooled steadily as the Universe expanded. But, as we shall see, that background radiation, the echo of creation, had a key role to play in persuading cosmologists that one of their 'model universes' might actually be telling them something deeply significant about the real Universe. And all this was happening while the person who was to become a key player in taking cosmology that step further in the 1970s, back to the beginning itself, was experiencing upheavals of his own, both personal and professional.

6

Marriage and Fellowship

The mid-sixties turned out to be one of the most important times in Stephen Hawking's life. Having become engaged to Jane, he realized that he would need to find a job very quickly if they were to be married. After obtaining a doctorate, the next stage in the career of any academic is usually to secure a fellowship, accompanied by a grant, in order to continue research. Much like the transition from undergraduate studies to postgraduate research, applications for fellowships are usually made while working on a PhD, rather than leaving things until afterwards. So, while in the throes of writing up his thesis, and with a wedding planned for the coming summer, Hawking had to look around for available posts. Fortunately he did not have to look far. He heard about a theoretical physics fellowship being offered by another college at the University, Caius,* to begin that autumn. Without hesitating he began to organize his application. However, getting such a relatively simple thing off the ground did not turn out to be as easy as he had hoped.

At this stage of his illness he was unable to write, and had planned to ask Jane to type his application during her next visit to Cambridge the coming weekend. But when his fiancée stepped off the train, she greeted him with her arm in plaster up to the elbow. She had had an accident the previous week

* Pronounced 'keys'; its full name is Gonville and Caius College.

and broken her arm. Hawking admits that he was not as sympathetic towards Jane as perhaps he should have been when he saw the state she was in, but hurt feelings were quickly mended and together they tried to work out how they could get the application written. Jane's left arm had been broken and she is right-handed, so Hawking dictated the information and she was able to write the application by hand. They managed to get a friend in Cambridge to type it up for them.

However, that was not the end of Hawking's problems. As a requirement of the application he had to give two references. Obviously Dennis Sciama was his first referee; he was, naturally, very supportive, and suggested Hermann Bondi as the second. Hawking had met Bondi on several occasions at the King's College seminars given by Roger Penrose earlier that year, and Bondi had communicated to him a paper he had written to The Royal Society a few months earlier. Encouraged by this, Hawking decided, with near-catastrophic consequences, to ask Bondi to give him a reference. As Hawking puts it:

I asked him after a lecture he gave in Cambridge. He looked at me in a vague way, and said, yes he would. Obviously, he didn't remember me, for when the College wrote to him for a reference, he replied that he had not heard of me.[1]

If such a serious blow had happened today he would almost certainly not have had a hope of getting his fellowship. In the sixties, however, competition for academic posts was not quite as fierce as it is now, and the authorities at Caius showed great tolerance in writing to tell him of the embarrassing situation. Sciama came to the rescue again, and contacted Bondi to refresh his memory about the promising young researcher. Bondi then gave Hawking a glowing reference, possibly far kinder than one he might originally have written.

The College Council at Caius meets annually during the Lent term to elect new fellows. There are usually six or seven positions on offer, covering the full spectrum of subjects, and if elected the successful applicant joins the seventy-odd fellows already in residence at the College. The Council consists of around a dozen senior fellows, headed by the College Master. In 1965 the Master was the famous historian of Chinese science, Joseph Needham. Hawking came with good recommendations, and a number of the science fellows on the council, including Needham, had heard of him via the early reputation he had already gained in Cambridge academic circles. As Shakespeare says, 'Sweet are the uses of adversity,' and maybe this has never been truer than in Hawking's case. Despite the confusion over references, the Council favoured him over his competitors, and he received his fellowship at Caius. As far as Hawking's career was concerned, he and Jane could now look to the future with a degree of confidence.

The duties of fellows are minimal beyond the basic condition that they continue with their research. They are required to do a little student supervision, but the level to which this is taken varies enormously. The role of the fellow, like many other things at Cambridge University, has changed little since Sir Isaac Newton's time. Fellowship is considered a great honour and a means by which academics may continue with their research and be paid for it. In return, a College gains prestige if one of its fellows turns out to be highly successful.

Possessing more than his fair share of cheek, Hawking nearly blew it again after having secured his fellowship at Caius. He managed this feat by almost pushing things too far with the Bursar. On a whim, he decided to ask him what he would be paid for his new position, and was rebuked for his impertinence. Although he could not foresee it at the time, soon after they were married this *faux pas* would cause him and Jane still further problems.

The couple were married in July 1965 in the chapel of

Hawking's postgraduate college, Trinity Hall. It was not a typical 'academic' wedding, but neither was it, by any means, a society occasion. Both sets of parents were ordinary, middle-class people. Jane's father, George Wilde, was a civil servant, and the Wilde family had known the Hawkings for some time before their children had met, so the wedding arrangements were perhaps a little less fraught with arguments than they might have been. Around a hundred guests attended, and the service was followed by a reception with all the usual speeches and champagne toasts to the happy couple. Brandon Carter remembers the wedding as the first occasion on which he met the Hawking family. He recalls Frank Hawking as a tall, slim man with a quiet and dignified air about him. Hawking's mother Isobel was instantly friendly and chirpy, a lively, gregarious character who delighted in meeting Stephen's friends and accepting them into the fold.

Despite the fact that the groom had to lean on a cane for the wedding photographs, the couple looked much the same as any other on their wedding day. In the black-and-white photographs Hawking is wearing a dark suit and a thin, neatly knotted tie, his dark-rimmed glasses and thin face giving him an owlish look. Jane stands beside him, hands clutching a bouquet of flowers, her veil pushed back to reveal shoulder-length hair curled outwards above the neckline of her short wedding dress in the fashion of the day. Hawking looks at the camera with a proud expression, a stare of deep-rooted determination and ambition – a stance that says, 'This is just the beginning.' Jane smiles happily at the lens, equally sure, in her own gentler way, that they will make out and overcome all adversity.

Of course they both knew, as did all the others on that day, that Stephen might die within a short time. In fact, according to the medical predictions he was already living on borrowed time. But such thoughts were only a distant shadow that summer's day in Cambridge, and Jane and Stephen Hawking were as sure as any other newly married couple that they would create a successful and happy life for themselves, and that because of their circumstances they would make the very most of every moment they had together.

A fellow's salary is no princely sum, and in 1965, foreign holidays were still relatively unusual, so the newly-weds honeymooned in Suffolk for a week. Immediately afterwards it was back to work, because the couple had to leave for a summer school in general relativity that Hawking was due to attend at Cornell University, in upstate New York. Hawking recalls that this was a mistake:

> It put quite a strain on our marriage, especially as we stayed in a dormitory that was full of couples with noisy small children. However, the summer school was very useful for me because I met many of the leading people in the field.[2]

Brandon Carter attended the same summer school and got to know Jane much better than he had during her weekend visits to Cambridge. He remembers that she was rather inexperienced at the traditional tasks of a housewife. He recalls how, on one occasion, he came across her in the shared kitchen practically pulling her hair out trying to make tea without a teapot. Carter found a saucepan in a cupboard and showed her how to brew tea camping-style. One of the fondest memories he has of that summer school is the look of indignation on Jane's face.

The idea of a summer school is to introduce the latest ideas to research students and fellows from universities around the world. They are usually attended by the most eminent people in a given field, and help to set scientists thinking about how to apply new discoveries to their own work. Hawking was getting into his stride as a physicist at this point in his career and, despite domestic difficulties, it was perfect timing as far as his cosmological ideas were concerned. He returned inspired to Caius and his first job.

However, upon their return there was a whole new set of domestic problems to face. The first of these was the matter of where the Hawkings were to live. Jane was still a student in her third and final year at Westfield College in London, so the plan was for her to stay in London during the week, while Stephen looked after himself, and, just as in the days before their marriage, she would return at weekends. The immediate problem was to find suitable accommodation in a university city where accommodation was always at a premium.

Before leaving for America, Hawking had gone to see the Bursar again to ask for assistance in finding somewhere to live, only to be told that it was against college policy to help fellows with housing. Because Stephen could not use a bicycle and was only able to walk short distances assisted by a pair of sticks, it was, of course, essential for the Hawkings to live in central Cambridge, close to the Department of Applied Mathematics and Theoretical Physics in Silver Street. But as far as the college authorities were concerned, their latest fellow's disabilities made no difference. Then, just before the trip to Cornell, they had heard of a new block of flats being built a short distance from the DAMTP and had put their names down for an apartment there. When they arrived back in Cambridge the Hawkings discovered that the flats would not be ready for several months.

In desperation, Hawking went back to the Bursar, who finally made the concession of arranging for the couple to stay in rooms in a hostel for graduate students. It seems, however, that the Bursar was still smarting over Hawking's check in asking what he would be paid for his fellowship. The normal price for a room was twelve shillings and six pence (63p) per night, but he charged the Hawkings double because there were two of them, even though Jane planned to stay there only at weekends.

In the event, they stayed in the hostel for just three nights because they discovered that a small house had become available nearby in a tiny street of picturesque old houses called Little St Mary's Lane. Less than a hundred yards from the DAMTP, it suited them perfectly. The house was owned by one of the other Cambridge colleges, which had let it to one its own fellows. He had now bought and moved into a larger house in the suburbs, and agreed to sublet the property for the remaining three months of his lease.

During their stay there they heard news of another house that had become available in the same street. An elderly neighbour who had befriended the couple discovered their housing problems, and contacted the owner of the empty house just a few doors along Little St Mary's Lane. Incensed by the idea that a struggling young couple should have such problems when a house remained unoccupied only a few yards away, the neighbour summoned the owner to Cambridge and insisted that the house be rented to the Hawkings and at a reasonable price. Once again problems had been turned on their head. They moved in when the three-month contract for the first house had run its course and were to remain there for many years.

The actual process of moving house was quite a problem, even if it was only a few doors along the same street. Their friends all mucked in, carrying furniture along the pavement and arranging it in the new place while Stephen leaned on his sticks, giving instructions and acting the part of foreman, shouting orders in his best coxswain's voice. Brandon Carter and Martin Rees both lent a hand, as did another friend, Bob Donovan, a chemistry postgraduate who had made friends with Stephen and Jane before their marriage.

The new house was another tiny, ancient building. The front door opened directly into a sitting-room, and there was a kitchen at the rear. A winding narrow staircase led up to the master bedroom on the first floor; beyond that, on the second floor, were a couple of smaller rooms. The Hawkings had very little furniture, and a large dining-table took up most of the space in the sitting-room. The walls were painted in soft shades; bright prints were hung around the room to give a splash of colour between sets of shelves lined with rows of books and records. The ceilings were low, and tall visitors had to crouch under doorways to avoid a bump on the head.

The Hawkings have always been enthusiastic hosts, and the tiny house was frequently crowded with friends who would come for supper or lunch at weekends, all gathered around the dining-table, trying to avoid talking shop but not always succeeding. Brandon Carter remembers the house in Little St Mary's Lane as a very cheerful place, where friends would all help out with the preparation of meals and the washing-up, the strains of Wagner or Mahler playing in the background.

Meanwhile, Hawking's work on black holes was progressing well. In December 1965, he was invited to give a talk at a relativity meeting in Miami. Jane was on her Christmas vacation from Westfield College, and although she was working towards her Finals that coming summer she decided to go to America with her husband.

By the time of the Miami meeting, Hawking's speech had deteriorated to a severe slur, and he was concerned that the audience would find it difficult to understand him. Fortunately one of his old friends, George Ellis, was spending a year at the University of Texas at Austin and would also be attending the Miami meeting. After a discussion in their hotel room, it was agreed that Ellis would give the talk on Hawking's behalf. It was a resounding success and, with the ink still wet on his PhD diploma, his work on singularity theory was enthusiastically received by some of the most eminent scientists gathered there from all over the world.

In Miami they stayed at the Fountainbleau Hotel, which had recently been used in the filming of the James Bond movie *Goldfinger*. It was a large hotel with a private beach. On one of their free days during the conference, George Ellis and his new wife spent the afternoon on the beach with Stephen and Jane. Around six o'clock in the evening, the spectacular red disc of the sun low in the west, they decided to return to the hotel for supper, only to find that the beach gates had been locked. A quick search for a way off the beach showed them that the only way they could get back into the hotel was through an open kitchen window at the side of the building. The problem was, how on earth were they to get Stephen, who could not even walk without the aid of sticks, through the window and back to their rooms?

They managed somehow to clamber through the opening and were halfway to getting Stephen through when they discovered that they were being watched by some Hispanic cleaners, who were not exactly pleased to see a weird-looking group of people struggling to get what looked like a lifeless body in through the kitchen window. Never were the Hawkings and the Ellises more thankful that Jane was studying modern languages. As soon as she realized the nationality of the cleaners, she began to talk to them in fluent Spanish and rapidly explained their predicament. Once they understood what was going on they were entirely hospitable, helped to get Stephen into the hotel kitchen and even guided the foursome back to their rooms.

George Ellis invited the Hawkings to stay in Texas for a short holiday. Jane was not due back in London until January, so they decided to go along. They spent a week in Texas, sightseeing and relaxing after a tiring term in their respective careers. The four of them went on long drives in the Ellises' car through the dramatic, rugged Texas landscape, drinking cold beers at remote desert bars and window-shopping in the Austin shopping malls.

Upon their return to Cambridge, the realities of life hit

them hard. Jane had to return almost immediately to London, and the old system of weekend visits began again.

During the first year of their marriage Jane really came into her own. She managed to continue with her studies and graduated in the summer of 1966. During that time she also typed up Stephen's PhD thesis, and continued to travel back to Cambridge every weekend and during holidays. In the summer of 1966, she was at last able to live with her husband throughout the week in their home in Little St Mary's Lane.

Meanwhile, Stephen's condition had begun to worsen. The nature of the disease is such that in many cases it progresses in irregular leaps. A period of little change, which may last for years, may be followed by a rapid decline and then a levelling-off. Since his diagnosis and early deterioration, Hawking's symptoms had remained more or less constant, but in the latter half of the 1960s another rapid decline occurred. He had to take to using crutches rather than sticks in order to get around. At this point his father became disillusioned and impatient with the advice his son was receiving from the medical profession and decided to take over Stephen's treatment. He carried out intensive research into ALS and prescribed a course of steroids and vitamins which Stephen continued to take until his father's death in 1986.

He was finding it increasingly difficult to negotiate the winding staircase to their bedroom on the first floor at Little St Mary's Lane. Friends who visited the couple for the evening began to appreciate just how much Stephen's condition had deteriorated as they saw him struggling across the sitting-room and up the stairs when he decided to retire for the night. One acquaintance has recalled that he watched in shock as Hawking took a full fifteen minutes to make the journey from the first stair to his bedroom door. He would never allow himself to be helped on these occasions, and utterly rejected any behaviour that singled him out as anything other than a normal, able-bodied man. Jane and their friends respected this attitude, but it could become frustrating at times. Hawking's determination and single-mindedness could often be misconstrued as arrogance and bloody-mindedness. The writer John Boslough has described Hawking as 'the toughest man I have ever met'.[3] And Jane has said, 'Some would call his attitude determination, some obstinacy. I've called it both at one time or another. I suppose that's what keeps him going.'[4]

At the DAMTP, and in Cambridge academic circles, Hawking was beginning to cultivate a 'difficult genius' image, and his reputation as successor to Einstein, although embryonic, was already beginning to follow him around. People who knew him in those days remember him as a friendly and cheerful character, but already his natural brashness, coupled with his physical disabilities, was beginning to create communication difficulties with many of those around him.

He was quite outspoken when attending talks given by internationally famous and highly respected figures in the world of physics. Where most young researchers would be happy to accept the words of the great quietly, Hawking would ask deep, often embarrassingly penetrating questions. Instead of alienating him from his seniors, this behaviour, quite rightly, gained him a great deal of respect and helped to increase his standing in the eyes of his superiors. However, it could be quite intimidating to some of his contemporaries. On occasion some colleagues felt a little shy about asking him to go for a beer at the pub.

Hawking's great personal gift is to be able to make light of his disabilities and always to have a cheerful and positive outlook on life. He simply refuses to let his condition get him down. In physics he has a perfect displacement activity. By keeping himself totally preoccupied with the nature and origin of the cosmos, and playing what he calls 'the game of Universe', he does not allow himself to spend time and energy thinking about his state of health. Once, when asked whether he ever became depressed over his condition, he replied, 'Not normally. I have managed to do what I wanted to do despite it, and that gives a feeling of achievement.'[5] Despite the gradual deterioration in his speech and increasing muscular atrophy, to his close friends he was the same Stephen Hawking they had known since his early Cambridge days, and those who really understood him felt the warmth of his personality.

Both Jane and Stephen knew that they should not waste any time in starting a family once they were married, and their first child, a boy they named Robert, was born in 1967.

This event was another turning-point in Hawking's life. Only four years after he was diagnosed as having a terminal illness and a life expectancy of two years, his reputation as a physicist was in the ascendant; he had retained, by sheer determination and will-power, a degree of independence and mobility; and now, against all odds, he was a father. As Jane has observed, 'It obviously gave Stephen a great new impetus, being responsible for this tiny creature.'[6] Everything seemed to be going well for him. His career was blossoming, and with every new paper he published a further barrier in our understanding of the Universe was broken down. His reputation as a promising new name in the world of physics was reinforced with each fresh breakthrough. And now he had a son to add to the happiness of his married life.

For Jane, these events were not quite so elevating. To her fell the burden of raising a child, keeping the home together, and caring for a severely disabled husband who could do nothing to help her. She is quoted as saying:

When I married him I knew there was not going to be the possibility of my having a career, that our household could only accommodate one career and that had to be Stephen's. Nevertheless, I have to say I found it very difficult and very frustrating in those early years. I felt myself very much the household drudge, and Stephen was getting all the glittering prizes.[7]

On another occasion she said:

I can imagine how frustrating it must be for some physicists' wives when they expect help from able-bodied husbands that is not forthcoming. I have no illusions on that score, so it doesn't trouble me unduly.[8]

However, it would be many years before the inevitable tensions that were brewing would break to the surface.

The couple decided to buy the house in Little St Mary's Lane. Hawking swallowed his pride and returned to the Bursary at Caius to ask the College for a mortgage. They conducted a survey of the property, decided that it would not be a sound investment and turned him down. Once again, his status as a fellow was opening up very few 'real life' privileges. Undeterred, they went to a building society for the loan and were granted a mortgage. Stephen's parents gave them the money to do up the house, and the usual gang of friends once more helped out, this time with wallpapering and painting.

Although the house was small, they remained there for a number of years until, in the mid-seventies, it became too cramped for the growing family. But in the meantime it served their purposes as well as it had ever done. Newly decorated, it was even cosier than it had been as a rented property, and – what was more important – it was now their own home, providing a secure environment in which they could begin to raise a family.

The sixties were a great time to be alive and young. They were a time of tremendous, although in some ways misplaced, hope, an era of re-awakening two decades on from the end of the Second World War and all the privations that followed, a time of fresh beginnings and optimism in all spheres of life. The second half of the decade heralded the first real counter-cultural revolution in the West, bringing with it new music, new art and new literature. A few years earlier, the trial surrounding the censorship of D.H. Lawrence's *Lady Chatterley's Lover* had seen the dam of élitism and Victorian morality burst wide open with the immortal question, 'Is it a book you would wish your wife or your servant to read?' The Beatles, the Rolling Stones and, so it seemed, half the youth of Britain and America were experimenting with psychedelic drugs; dresses were getting shorter, and hair longer.

The Hawkings and their friends in Cambridge showed little interest in fashion and pop music, although Jane was keen on mini-dresses and the latest hair-styles. But in the world of science things were also on the move. George Ellis clearly remembers watching the maiden flight of the British Concorde, 002, in April 1969, and being filled with excitement at the new technology taking the world by storm. Then, only a few months later, they sat glued to their TV screens to watch the 'one small step' of Neil Armstrong when the lunar module, Eagle, landed in the Sea of Tranquillity, 240,000 miles away on the surface of the Moon. 'The Eagle has landed,' he said. 'The surface is like a fine powder. It has a soft beauty all its own, like some desert in the United States.' At that moment, anything seemed possible.

The Hawkings and the Ellises went on holiday together in 1969. Foreign holidays were suddenly in vogue because of drastically reduced prices, and it had become very fashionable to take a package trip to such destinations as Spain or its outlying islands, especially Majorca. The two families flew to Palma airport, Majorca, and spent a short break walking through the unspoilt almond groves, sampling the local wine and sunning themselves on the clean, unmolested beaches, almost untouched by visiting Anglo-Saxons and certainly lager-lout-free.

Hawking was working harder than he had ever worked before, and it was paying dividends. In 1966 he won the Adams Prize for an essay entitled 'Singularities and the

Geometry of Spacetime'. Much of his research during this period was a continuation of the work that had yielded the astonishing last chapter of his PhD thesis. He spent most of this time in collaboration with Roger Penrose, who was by then Professor of Applied Mathematics at Birkbeck College in London.

One of the major difficulties the two of them faced was that they had to devise new mathematical techniques in order to carry out the calculations necessary to verify their theories – to make them empirically sound and not just ideas. Einstein had experienced a similar problem fifty years earlier with the mathematics of general relativity. He, like Hawking, was not a particularly brilliant mathematician. Fortunately for Hawking, however, Penrose was. In fact, he was fundamentally a mathematician rather than a physicist, but at the deep level at which the two subjects become almost indistinguishable.

It really boils down to a difference in approach. Hawking's way of working is largely intuitive – he just knows if an idea is correct or not. He has an amazing feel for the subject, a bit like a musician playing by ear. Penrose thinks and works in a different way, more like a concert pianist following a musical score. The two approaches meshed perfectly, and soon began to produce some very interesting results on the nature of the early Universe. As Dennis Sciama has put it, '[The theories] required very highbrow methods, at least by the standards of theoretical physicists.'[9] Penrose liked to work in a highly visual way, using diagrams and pictures, which suited Hawking fine. He always felt more at home with visual representations than with mathematical formulae. It was also so much easier for him to manipulate these pictures rather than trying to work with equations which he could not write out and had to retain in his head.

Since his undergraduate days Hawking has been a keen follower of the philosopher Karl Popper. The main thrust of Popper's philosophy of science is that the traditional approach to the subject, 'the scientific method' as originally espoused by the likes of Newton and Galileo, is in fact inadequate.

The traditional approach to science can be broken down into six stages. First comes an observation or an experiment. Scientists then try to devise a general theory to explain by induction what they have observed, and go on to propose a hypothesis based on this general theory. Next come attempts to verify this hypothesis by further experimentation. The original theory is thus proved or disproved, and the scientist then assumes the truth or otherwise of the matter until proven wrong.

Popper stands this process on its head, and suggests the following approach. Take a problem. Propose a solution or a theory to explain what is happening. Work out what testable propositions you can deduce from your theory. Carry out tests or experiments on these deductions in order not to prove them, but to refute them. The refutations, combined with the original theory, will yield a better one.

The primary difference between the two approaches is that, according to the traditional scientific method, after making an observation the scientist attempts to verify a theory by further experiment. In Popper's system, the scientist tries to disprove the theory in an attempt to find a better one. It is this aspect of Popper's thought that is so appealing to Hawking and many other scientists, and he has often applied it in his own scientific work. The science writer Dennis Overbye once asked him how his mind worked. In reply, Hawking said:

> Sometimes I make a conjecture and then try to prove it. Many times, in trying to prove it, I find a counter-example, then I have to change my conjecture. Sometimes it is something that other people have made attempts on. I find that many papers are obscure and I simply don't understand them. So, I have to try to translate them into my own way of thinking. Many times I have an idea and start working on a paper and then I will realize halfway through that there's a lot more to it.

I work very much on intuition, thinking that, well, a certain idea ought to be right. Then I try to prove it. Sometimes I find I'm wrong. Sometimes I find that the original idea was wrong, but that leads to new ideas. I find it a great help to discuss my ideas with other people. Even if they don't contribute anything, just having to explain it to someone else helps me sort it out for myself.[10]

Little did he know, at the end of the 1960s, just how important his ideas would soon prove to be.

7

Singular Solutions

During the 1960s, four new developments, two concerning black holes and two cosmological, led to a revival of interest in the singular solutions to Einstein's equations. As a result of the work stimulated by these developments, especially the collaboration between Hawking and Roger Penrose, physicists realized at the beginning of the 1970s that they might have to come to terms with the unthinkable: the prediction from the general theory of relativity that points of infinite density – singularities – *could* exist in the Universe did not, after all, indicate a flaw in those equations, and singularities might *really* exist. Even worse, for those still trying to cling to an older picture of reality, because the Universe itself seems to be a black hole viewed from within the Schwarzschild horizon, there might indeed be a singularity at the beginning of time that could *not* be obscured from our view – a 'naked' singularity.

It all began with the discovery of quasars, in 1963. The quasar story actually began on the last day of 1960. During the 1950s, astronomers using telescopes sensitive to radio waves rather than visible light had identified many objects in the Universe that produce a lot of radio noise. Some of these objects were also visible as bright galaxies, and were known as radio galaxies, but others had not yet been identified with any known visible object. Then, at the end of 1960, the American astronomer Allan Sandage reported that one of the radio sources discovered during a survey carried out by radio astronomers in Cambridge, England (and known as 3C 48) could be identified not with a distant galaxy, but with what seemed to be a bright star. More of these radio 'stars' were identified over the next few years, but nobody could explain how they produced the radio noise. Then, in 1963, Maarten Schmidt, working at the Mount Palomar Observatory in California, explained why another of these objects, known as 3C 273, had a very unusual spectrum.

All stars (and other hot objects) reveal their composition by the nature of the light they emit. Each kind of atom, such as hydrogen, helium or oxygen, absorbs or emits energy only at very precise wavelengths, because of the quantum effects mentioned in Chapter 2. So when light from a star or galaxy is spread out, using a prism, into a spectrum, we see that the spectrum is crossed by a series of dark and bright lines at different wavelengths, corresponding to the presence of atoms of different elements in the atmosphere of the star (or in the stars that make up the galaxy). These spectral lines are as characteristic as fingerprints, and for a particular type of atom they are always produced at the same distinctive wavelengths.

Astronomers already knew, though, that these spectral lines are shifted a little bit towards the red end of the spectrum in the light from galaxies outside the Milky Way. This famous 'redshift' is caused by the expansion of the Universe, which stretches space, and therefore stretches the wavelength of light *en route* to us from a distant galaxy. Indeed, it was the discovery of the redshift that told astronomers the Universe must be expanding, just as Einstein's equations had predicted, but Einstein himself had at first refused to believe it.

The fact that light from 3C 273 was redshifted – the discovery Maarten Schmidt made – was not a surprise; but the size of the shift, nearly 16 per cent towards the red end of the spectrum, astonished astronomers in 1963. Typical redshifts for galaxies are much less than this, about 1 per cent, or 0.01. With the realization that such large redshifts were possible, other radio 'stars' were re-examined, and it turned out that they all showed similar or even larger shifts. 3C 48, for example, has a redshift of 0.368 (nearly 37 per cent), more than twice that of 3C 273, and the record redshift now stands above 4 (in other words, the light from the most distant quasars known is stretched to more than four times its original wavelength).

In the expanding Universe, redshift is a measure of distance (the farther light has to travel on its way to us, the more it will be stretched by Universal expansion). So these objects were not stars at all, but something previously unknown – objects that looked like stars but were far away, in most cases farther away than the known galaxies. They soon became known as quasistellar objects, or 'quasars'.

In order to be visible at all at the huge distances implied by their redshifts, quasars must produce prodigious amounts of energy. A typical quasar shines with the brightness of three hundred billion stars like the Sun, three times as bright as our whole Milky Way Galaxy. Having sought in vain to find any alternative means to explain the power of quasars, astronomers were reluctantly forced to consider the possibility that they might be black holes. We now know that each quasar is a black hole containing at least a hundred million times as much mass as our Sun, contained within a volume of space with about the same diameter as our Solar System. (This is just the kind of large, low-density black hole described in Chapter 5.) Each one actually lies at the heart of an ordinary galaxy, and feeds off the stellar material of the galaxy itself. Ever-improving telescope technology has enabled us, in many cases, to photograph the surrounding galaxy itself, faint alongside the quasar.

Although a hundred million solar masses is large by

everyday standards, this still represents only one-tenth of 1 per cent of the mass of the parent galaxy in which a quasar lurks. When such an object swallows matter, as much as half the mass of the matter can be converted into energy, in line with Einstein's famous equation $E = mc^2$. As we saw in Chapter 5, the factor c^2 is so huge that this corresponds to a vast amount of energy. This process of energy production is so efficient that even if only about 10 per cent of the infalling mass is actually converted into energy, a quasar can shine as brightly as three hundred billion Suns, bright enough to be seen across the vast reaches of intergalactic space, if it is swallowing just one or two solar masses of material every year. The material forms a great, hot, swirling disc around the black hole itself. This disc is where the energy that produces the radio noise, and the visible light, comes from, even though the hole itself, as the name implies, is black. And with a hundred billion stars to eat, even if a quasar only dines off 1 per cent of the mass of the parent galaxy, it can shine that brightly for a billion years.

The existence of quasars shows that large, low-density black holes really do exist. In 1967, just four years after the redshift of 3C 273 was measured, the Cambridge radio astronomers achieved another breakthrough with the discovery of the rapidly varying radio sources that became known as 'pulsars'. And although pulsars are not themselves black holes, they opened the eyes of most astronomers to the possibility that superdense, compact black holes might also really exist, just as the general theory of relativity predicted.

The first pulsars were discovered by a research student, Jocelyn Bell, while testing a new radio telescope. The astonishing thing about these radio sources is that they flick on and off several times a second (some of them several *hundred* times a second) with exquisite precision. This is so much like an artificial signal, a kind of cosmic metronome, that, only half-jokingly, the first pulsars discovered were labelled 'LGM 1' and 'LGM 2' – the initials 'LGM' stood for 'Little Green Man'. As more of them were discovered, though, it became clear that there were far too many to be explained as interstellar traffic beacons set up by some super-civilization, and the accepted name became pulsar, from a contraction of 'pulsating radio source' and because the name chimed with quasar.

But what natural phenomenon could produce such regular, rapid pulses of radio noise? There were only two possibilities. The pulses had to signal either the rotation or the vibration of a very compact star. Anything bigger than a white dwarf would certainly rotate or vibrate too slowly to explain the speed of the known pulsars, and rotating white dwarfs were soon ruled out – a simple calculation showed that a white dwarf rotating that fast would break apart.

For a short time early in 1968, it seemed that vibrations of a white dwarf, literally pulsing in and out, might explain the variations in the radio noise from pulsars. But it was fairly straightforward to calculate the maximum rate at which a white dwarf could pulsate without breaking apart. Indeed, one of us (J.G.) did exactly that as part of the work for his PhD. The answer was disappointing (for him) but conclusive: white dwarfs simply cannot pulsate at the required speed. Which meant that the stars involved in the pulsar phenomenon must be even more compact, and more dense, than white dwarfs.

They must, in short, be neutron stars, predicted by theory, but never previously discovered. Within months of the announcement of the discovery of pulsars, it was established that these objects are actually rotating neutron stars, definitely within our Galaxy, producing beams of radio noise which sweep past us like the flashing beams of a lighthouse. They are created by the explosions of giant stars as supernovae. And, as theorists were well aware from the outset, the same theory that predicted the existence of neutron stars, a prediction which had been largely ignored for thirty-odd years, also predicted that by adding just a little more mass to a neutron star (or by having a little more debris left over from a supernova explosion) you would create a collapsar.

It is no coincidence that John Wheeler coined the term 'black hole' in this connection the year following the discovery of pulsars, for the realization that pulsars must be neutron stars triggered an explosion of interest in the even more exotic predictions of the general theory of relativity. That explosion had already been primed, however, by yet another discovery made using radio telescopes, which had confirmed the reality of the Big Bang itself.

When the Universe was more compressed, it was hotter, just as the air in a bicycle pump gets hot when it is compressed. The Big Bang was a fireball of radiation in which matter initially played an insignificant role. But as the Universe expanded and cooled, the radiation faded away, and matter, in the form of stars and galaxies, came to dominate the scene.

All this was known to astronomers in the 1940s and 1950s. George Gamow and his colleagues even carried out a rough calculation of what temperature this left-over radiation would have cooled to by now. In 1948, they came up with a figure of about 5 K (*minus* 268°C). By 1952, Gamow was inclined to think that it might be rather higher, and in his book *The Creation of the Universe* he said that the temperature ought to be somewhere below 50 K. But 5 K or 50 K, it was still a very low temperature, and in the 1950s nobody seriously contemplated the possibility of trying to detect this echo of creation, a cold background sea of radiation filling the entire Universe, and left over from the Big Bang.

In the early 1960s, though, the possibility of actually measuring the strength of this background radiation, and thereby testing the Big Bang model, occurred to a few astronomers. One way of understanding how and why the radiation has cooled is in terms of redshift. Radiation that filled the Universe in the Big Bang still fills the Universe, but because space has stretched since then the waves making up that radiation have had to stretch accordingly in order to fill the space available. This means that energy that started out in the form of X-rays and gamma-rays would now be in the form of microwaves, with wavelengths of around 1 millimetre or so. These are just the kind of radio waves used in some communications links, and in radar. With the technology developed for radar and radio communications, and the associated rapid development of radio astronomy, researchers in both the Soviet Union and the United States saw that the background radiation predicted by the Big Bang model might be detectable, and set about designing and building radio telescopes to do the job.

But they started just too late. The American team, based at Princeton University, was headed by Robert Dicke, who had worked in radar during the Second World War. In the early 1960s he set a team of young researchers the task of building a microwave background detector using an updated version of equipment he had helped to design during the war. By 1965 things were progressing nicely, when Dicke received a phone call from a young researcher at the Bell Laboratories, just 30 miles away from Princeton. The caller, Arno Penzias, wanted Dicke's advice about some puzzling radio interference that Penzias and his colleague Robert Wilson had been getting on their radio telescope at Bell Labs since the middle of 1964.

Penzias and Wilson had, in fact, been using an antenna designed for use with the early communications satellites, modified to operate as a radio telescope. They found that, wherever they pointed the telescope in the sky, they seemed to be getting a signal corresponding to microwave radiation with a temperature of just under 3 K. After trying everything they could to sort out what was wrong with the telescope (including cleaning pigeon droppings off the antenna, in case that was what was causing the interference), they gave up and called Dicke, an expert on microwaves, to ask if he had any idea what was going on.

Dicke soon realized that Penzias and Wilson had, in fact, detected the background radiation left over from the Big Bang. The Princeton detector, completed hurriedly a little later, confirmed the discovery, and soon radio astronomers around the world were getting in on the act. We now know that the Universe is indeed filled with a weak hiss of microwave background radiation, with wavelengths of around 1 millimetre, corresponding to a temperature of 2.73 K.

It was this discovery that opened the eyes of cosmologists to the reality of the Big Bang model: *not* just a model, after all, but an accurate description of the real Universe we live in. First, the existence of the background radiation showed that there really had been a Big Bang; then, by using the precise measurement of the temperature of that radiation today, it was possible to work backwards to the Big Bang to calculate the exact temperature of the fireball itself. We got slightly ahead of our story in Chapter 5, when we described the first few minutes of the life of the Universe – the accuracy of that description, dating from the mid-1970s, depends in part on our present-day knowledge of the precise temperature of the background radiation. But there is something else significant about that description of the early stages of the Universe. *The First Three Minutes* was not written by a specialist in cosmology, or even by an astronomer, but by a mainstream physicist, Nobel prize-winner Steven Weinberg.

Before 1965, cosmology was a quiet backwater of science, almost a little ghetto where a few mathematicians could play with their models without annoying anybody else. Today, a quarter of a century later, the study of the Big Bang is at the centre of mainstream physics, and Big Bang cosmology is seen as offering the key to understanding the fundamental laws and forces by which the physical world operates. It is because of the measurements of the background radiation that we can be so confident about how nuclei were synthesized in the Big Bang. And it was the first calculations of this kind made after the discovery of the background radiation that convinced many physicists (not just cosmologists) that hot Big Bang cosmology had to be taken seriously as a description of the Universe.

These calculations were not something hurriedly cooked up in the light of the discovery of the background radiation, but represented the culmination of more than ten years' work. In the 1950s, inspired by Fred Hoyle's lead, a team of British and American researchers had worked out how all the elements more complex than helium are synthesized inside stars. This was an astonishing *tour de force*. In essence, the process consists of sticking helium-4 nuclei together to build up more complex nuclei. Some of the complex nuclei then either spit out or absorb the odd proton, making nuclei of other elements.

As we mentioned in Chapter 5, though, there is a bottleneck for this process at its earliest stage. There is no stable nucleus that can be made by sticking two helium-4 nuclei together, and that is why nucleosynthesis stopped with helium in the Big Bang. Hoyle found a way round this bottleneck, via extremely rare collisions of three helium-4 nuclei almost simultaneously. This makes it possible to create a nucleus of carbon-12, but only if the energies (speeds) of the helium-4 nuclei are just right. The energies are just right inside stars, thanks to an unusual quantum effect known as a resonance. Nobody realized this until Hoyle explained how the crucial step in the chain must take place. He predicted the existence of the crucial resonance, which was then found during experiments here on Earth. Together with his colleagues, Hoyle went on to explain how *everything* is built up from hydrogen and helium inside stars – including the atoms in your body, and in this book.

In one of the strangest decisions ever made by a Nobel committee, one of Hoyle's colleagues, Willy Fowler, later received a share of the 1983 Nobel Prize for Physics for this work. Fowler is a fine physicist in his own right, and was a key member of the team. But he is the first to acknowledge that Hoyle made the key breakthrough on carbon-12 production, and was the inspiration for the team's efforts. Unfortunately, later in his career Hoyle espoused some decidedly unconventional ideas about the possibility that outbreaks of disease on Earth might be caused by viruses from comets. It seems that the Nobel committee, in its wisdom (?), decided not to give him a share of the Physics prize with Fowler for fear of seeming to lend credence to what they regarded as his more cranky work. At least the British establishment, for once belying its stuffy image, acknowledged Hoyle's true worth with a knighthood. All that, however, lay far in the future in 1967, when Fowler, Hoyle and their colleague Robert Wagoner put the icing on the nucleosynthesis cake.

The one problem with the story of stellar nucleosynthesis as developed in the 1950s was that it could not explain where helium came from. Starting out with stars in which 75 per cent of the material was hydrogen and 25 per cent helium, the theory could explain beautifully the presence of every other element, and could even explain why some elements are more common than others, and how much more common. But it all starts with the triple-helium/carbon-12 resonance, and without that initial 25 per cent of helium stars would not be able to cook up the rest of the elements. It was Wagoner, Fowler and Hoyle who together showed that the kind of Big Bang that would leave a background radiation with a temperature of 2.73 K today would also produce a mixture of

25 per cent helium and 75 per cent hydrogen at the end of the first four minutes.

Their findings were unveiled at a meeting in Cambridge in 1967. One of us (J.G.) was present, as a very junior research student, somewhat in awe of the occasion. He clearly recalls the deep questions being asked at the meeting by another member of the audience, a slightly older but still junior researcher, who seemed to have a slight speech impediment, but whose words were listened to closely by the more eminent researchers on the platform. Stephen Hawking was already known to be someone worth listening to, even at this early stage of his career. And the reason for his keen interest in Big Bang cosmology soon became clear, when the results of the investigation he was carrying out with Roger Penrose were published.

Hawking had begun puzzling over the singularity at the beginning of time in the early 1960s, but had soon been deflected, as we have seen, by the diagnosis of his illness, temporarily giving up his work. But by 1965 things were looking up. He had decided that he wasn't going to die quite so quickly as the doctors had predicted, after all; he had met and married Jane; and he was back at work with a vengeance. He was one of the few people, at that time, to take seriously the more extreme predictions of the general theory of relativity. Two years after the identification of the first quasar (but before its energy source was explained), and two years before the discovery of pulsars, only a handful of people believed in the possibility that black holes might exist, or that the Universe really had been born out of a singularity.

One of the few other people who did take the notion of black holes seriously was a young mathematician, Roger Penrose, working at Birkbeck College in London. It was Penrose who showed that every black hole must contain a singularity, and that there is no way for material particles to slide past each other in the middle of the hole. Not just matter, but spacetime itself simply disappears at the singularity. At such a point the laws of physics break down, and it is impossible to predict what will happen next.

But, as we have seen, this need not be too worrying, provided such bizarre objects are always safely hidden behind the horizon of a black hole. In this spirit, Penrose proposed a 'cosmic censorship' hypothesis, suggesting that all singularities must be hidden in this way, and that 'nature abhors a naked singularity'. In other words, observers outside the horizon of the black hole are always protected from any consequences of the breakdown of the laws of physics at the singularity.

Hawking was intrigued by Penrose's work on singularities, but saw that there was no way nature's abhorrence of a singularity could shield us from the singularity at the beginning of time – *if* it existed. In 1965, the two of them joined forces to investigate this puzzle.

Previously, researchers had expected that if you tried to wind back the equations describing the expanding Universe, things would get more and more complicated as you approached the Big Bang. Particles would collide and bounce off one another, producing a chaotic and confusing fireball. To many people this looked like the ideal way to make a model universe bounce at high densities, without encountering a singularity. But, over the next few years, Hawking and Penrose developed a new mathematical technique for analysing the way that points in spacetime are related to one another. This did away with the confusion of the messy interactions between material particles, and highlighted the underlying significance of the expansion (or collapse) of space itself.

The end result of this study was their proof that there must have been a singularity at the beginning of time, if the general theory of relativity is the correct description of the Universe. There is no way for particles in a contracting universe to slide past one another and avoid meeting in a singularity in the fireball, any more than it is possible to avoid the singularity inside a black hole. After all, when space shrinks to zero volume, there is literally no room left for particles to slip past one another. In other words, the expansion of the Universe away from the singularity in the beginning really is the exact opposite of the collapse of matter (and spacetime) into a singularity inside a black hole. The cosmic censor has slipped up, and there is at least one naked singularity in the Universe that we are exposed to, even if it is separated from us by 15 billion years of time.

While Hawking and Penrose were working all this out, the discovery of the background radiation was announced, pulsars were discovered, and Wagoner, Fowler and Hoyle were explaining how helium had been made in the Big Bang. By the time the Hawking–Penrose theorems were published, John Wheeler had given astronomers the term 'black hole', and newspaper stories were being written about the phenomenon. What had started out as an esoteric (but erudite) piece of mathematical research had evolved by the end of the 1960s into a major contribution to one of the hottest topics in science at the time.

And yet, this was Hawking's *first* real piece of research, stemming from his PhD work – the journeyman piece for his scientific apprenticeship. What on earth would he come up with next? And what did it mean to say that there had been a definite beginning to time in the Big Bang? There seemed very little prospect, however, that the young researcher would come up with anything of comparable importance. The deterioration in his physical condition seemed to rule out a long career.

8

The Breakthrough Years

The 1960s ended with Hawking being forced to make a concession to his physical condition. After a great deal of persuasion from Jane and a number of close friends, he decided to abandon his crutches and take to a wheelchair. To those who had watched his gradual physical decline, this was seen as a major step and viewed with sadness. Hawking, however, refused to let it get him down. Although the acceptance of a wheelchair was a physical acknowledgement of his affliction, at the same time he gave it not the slightest emotional or mental endorsement. In every other way, life went on as usual. And he could not deny that it did enable him to get around more easily. Never giving in to the symptoms of ALS more than he is physically compelled to is all part of the Stephen Hawking approach to life. As Jane said, 'Stephen doesn't make any concessions to his illness, and I don't make any concessions to him.'[1] That seems to be the way he has survived against all the odds for so many years, and also how Jane managed to remain sane living with him.

Earlier, in 1968, Hawking had been invited to become a staff member at the Institute of Theoretical Astronomy housed in a modern building on the outskirts of Cambridge. Originally it was headed by Fred Hoyle, but he resigned his post in 1972 after a final blazing row with the Cambridge establishment. This time the dispute was over the administration of British science in general and Cambridge science in particular. When Hoyle left, the Institute was merged with the Cambridge Observatories and came under the control of Professor Donald Lynden-Bell. Under his leadership 'Theoretical' was dropped from the name, and it has been the Institute of Astronomy ever since. In the same year a young radio astronomer, Simon Mitton, was appointed administrative head of the Institute. He subsequently worked closely with Hawking during the years he spent there.

Hawking worked at the Institute three mornings a week. It was too far from Little St Mary's Lane to get to by wheelchair. Instead, he had managed to acquire a three-wheeled invalid car, which he drove out into the suburbs on the main roads. Mitton would meet him at his car and help him out of the little blue vehicle and into the main building. Hawking had his own office, and as his prestige grew during the following years, a string of eminent astronomers and theoretical physicists were drawn to the Institute to confer with him. Mitton describes him as a human magnet in the world of physics. Graduate students as well as professional scientists from all over the world were attracted to the Institute mainly because of his presence there.

Hawking was never interested in observational astronomy. While an undergraduate at Oxford, he had attended a vacation course at the Royal Greenwich Observatory, helping the then Astronomer Royal, Sir Richard Woolley, to measure the components of double stars. However, so the story goes, upon looking through the telescope and seeing nothing more impressive than a couple of hazy dots in the star-field, he was convinced that theoretical physics would be more interesting. To this day he has looked through a telescope no more than a handful of times. At the Institute of Astronomy the work Hawking was interested in pursuing was conducted in his head or with pen, paper and computer.

Mitton recalls that Hawking was not the easiest person to work with. He found him irritable and impatient, and he remembers very little of the famous Hawking wit and humour. Secretaries apparently also found him difficult, and there were many occasions when a newly employed assistant would come to see Mitton on the verge of tears, complaining of over-demanding work-loads. Hawking always wanted things done yesterday. At such times Mitton had to remind himself and the secretaries working for him that such moods were perhaps a symptom of the man's condition.

Others would disagree. Roger Penrose has pointed out that Hawking displays an unusual cheerfulness and sense of humour in the face of adversity. He has seen Hawking in a bad mood, irritable and impatient with those around him, but he believes that many people with ALS develop a compensation mechanism, a system which acts as an anti-depressant. It would perhaps be nearer the mark to say that Hawking's behaviour is more to do with his own character than any effect of his illness. Like the rest of us, he is sometimes short and impatient with those around him, and he does not suffer fools gladly. Because he works at such an intense pace, putting great demands on himself, he expects everyone else to have the same energy and drive. Perhaps he simply didn't get on with the secretaries at the Institute of Astronomy.

However, the Institute seemed to be more aware of his worth than his own college was. The authorities made every effort to assist him in his work and to compensate for his disabilities. They had an automatic phone fitted in his office, pre-programmed to enable him to reach other numbers at the push of a single button. But this was long before digital technology, and the device was really little more than a box of tricks with a vast number of leads and connections sprouting from a junction box in the corner of the room. It took Post-Office engineers over a week to install it.

There was a definite buzz in Cambridge about Hawking and his work, even before he joined the Institute of Theoretical

Astronomy. He had a certain aura about him. Long before he had made his mark on cosmology, among graduate students there was an air of reverence accompanying the name Stephen Hawking. Such early discipleship illustrates the beginnings of the cult status that has surrounded many of the things Hawking has said and done during his career. Even in the early 1970s, it was possible to see that the image of the crippled genius, so beloved of the media, was beginning to take root in the minds of those on the periphery of Hawking's life and work. Instead of this image diminishing or fading away as his career has blossomed, with each fresh achievement his status as the new Einstein, the purely cerebral creature trapped inside an inoperative body, has grown.

Mitton recalls that, by the time of their first meeting in 1972, Hawking's speech had deteriorated considerably. It was essential to concentrate hard on what he was saying in order to understand him. Mitton found that he always had to face Hawking and watch what he was saying as well as listening intently; even then it was not easy. The best way to communicate, Mitton found, was to ask questions which required only negative or affirmative answers. So, instead of asking, 'When would you like to go to lunch, Stephen?' it was far easier to say, 'We are going to lunch at 12.30, is that all right?' Fischer Dilke, who wrote and directed one of the first television documentaries about Hawking, disagrees. He says that Hawking hates nothing more than being asked such questions because it is a sign to him that the person he is talking to is not treating him in a normal way. It obliges him to answer only 'Yes' or 'No', and he would, quite naturally, like to be engaged in a standard conversation.

In retrospect the seventies may be viewed as something of a grey decade. After the optimism and hope of the sixties, the West, with the possible exception of West Germany, was thrown into recession; only in Japan did a combination of post-war determination, a flair for the commercial application of Western technology and sheer hard work set the pattern for industrial growth. Britain's economy nearly foundered, hammered by a series of disastrous strikes and political turmoil. The decade began with a Labour government, which lasted until June of 1970 when Edward Heath narrowly beat Harold Wilson in a surprise victory, and ended with a new style of Tory government in the shape of the country's first woman Prime Minister, Margaret Thatcher.

In April 1970 the world held its breath as the drama of Apollo 13 was enacted hundreds of thousands of miles out in space, and the crippled spaceship limped home to safety. In September high drama of a different kind was played out in the Jordanian desert when Middle Eastern terrorists blew up three jet airliners. The world lost a charismatic and influential figure in the shape of Hawking's schoolboy hero, Bertrand Russell, who died aged 97. And it was in that year that Stephen Hawking began to turn his attention towards the exotic astronomical objects recently dubbed 'black holes', and once again found himself in collaboration with the mathematician Roger Penrose.

It is often the case with scientific discovery that a crucial step forward comes through inspiration at an unexpected moment, and Hawking is fond of recalling the story of when his first black hole breakthrough came to him. Soon after the birth of his second child, Lucy, in November 1970, he was thinking about black holes as he got ready for bed one night. As he says:

> My disability makes this rather a slow process, so I had plenty of time. Suddenly, I realized that many of the techniques that Penrose and I had developed to prove singularities could be applied to black holes.[2]

At that time, notions of what a black hole was really like were pretty hazy, and both Penrose and Hawking had been trying to come up with some way of stating which points in spacetime were inside a black hole and which were outside. It was just as he was about to get into bed that an obvious solution struck him. The answer to the problem was actually one which he claims Penrose had originally suggested but had not applied to the situation they were studying. The science is described in the next chapter; suffice it to say that the resolution was so exciting that Hawking got very little sleep that night. Early the next morning he was on the phone to Penrose.

For the next two years (as we describe more fully in Chapter 9) the pair of them developed their ideas about the physics of black holes. As they worked, they came to see that the way they had originally perceived black-hole physics was not as clear-cut as it ought to be. To get to grips with it properly required them to dust away the mental cobwebs of dimly remembered physical concepts they had not thought about since undergraduate days. In particular, Hawking was gaining a renewed interest in a field called thermodynamics, developed by Lord Kelvin and others in the nineteenth century.

No one would have imagined that thermodynamics had any relevance to black holes at all. As Dennis Overbye has put it, 'It was as if he had popped the hood on a Ferrari and found an antique steam engine chugging away inside.'[3] It was ridiculous – thermodynamics was used to study gases under pressure, heat transfer and the efficiency of steam engines, not such exotic objects as black holes. Little did Hawking realize at the time that thermodynamics was to have a huge influence on the future of black hole theory and would shortly lead him into his second major scientific confrontation with another physicist.

By early 1973, Hawking and Penrose were beginning to use thermodynamics as an analogy for what was happening in a black hole. Scientists often do this: an everyday model helps them to understand situations as bizarre as those found in a singularity. However, a young researcher named Jacob Bekenstein, working at Princeton University, was taking things a lot further. He was not content to use thermodynamics as an analogy but instead was applying its precepts literally. And he was coming up with some very interesting results.

When Hawking discovered Bekenstein's work he was incensed. He had been using thermodynamics as nothing more than a model for what was going on, and believed it totally ridiculous to take it further and actually *apply* it to black holes. Together with his old friend from Cambridge, Brandon Carter, and the American relativist James Bardeen, he published a paper in the scientific journal *Communications in Mathematical Physics* which attempted to disclaim the suggestion. The argument raged in the scientific press and across the Atlantic for many months. Hawking was becoming more and more irritated by what he saw as Bekenstein's absurd notions. In reply to a paper published by Bekenstein, Hawking, Carter and Bardeen responded with their own, entitled 'The Four Laws of Black Hole Mechanics'. Both papers were later shown to be incomplete.

Most physicists sided with Hawking and his co-authors, but Bekenstein was not put off by the massed ranks of the scientific community ranged against him. Years later he said of the confrontation:

> In those days in 1973 when I was often told that I was headed the wrong way, I drew some comfort from Wheeler's opinion that 'black hole thermodynamics is crazy, perhaps crazy enough to work'.[4]

Hawking continued to think that Bekenstein's notion was simply crazy – at least for a while. What brought about the change was a series of events which would lead him to a far more important conclusion about black holes, and propel him to the forefront of theoretical physics. But that was half a year away, and in the intervening period the arguments continued.

Meanwhile Hawking was finding the mathematics of the work increasingly difficult to deal with. The equations for interpreting the physics of black holes are amazingly complex, and by this stage of his illness he could use neither paper and pen nor a typewriter. Instead, he was forced to develop techniques for keeping such information in his mind and ways of manipulating equations without being able to write them down. Such a feat has been described by one of Hawking's friends and collaborators, Werner Israel:

[The] achievement is as though Mozart had composed and carried an entire symphony in his head – anyone who saw the lines of complex mathematics covering the blackboard like musical staves at a recent seminar would have appreciated the comparison.[5]

Hawking has the great advantage of possessing a superb memory. In his book *Beyond the Black Hole: Stephen Hawking's Universe*, John Boslough recounts an incident which demonstrates Hawking's ability to retain detailed information in his head:

One of Hawking's students told me that, while driving him to London for a physics conference once, Hawking remembered the page number of a minor error he had read in a book years before.[6]

Another anecdote describes how a secretary who worked for Hawking was amazed when he had once recalled, twenty-four hours later, a tiny mistake he had made while dictating – from memory – forty pages of equations. Hawking is not unique in having this talent. In 1983, he dazzled students at a Caltech seminar when he dictated a forty-term version of an important equation from memory. As his assistant finished writing the last term, his colleague, Nobel laureate Murray Gell-Mann, who happened to be sitting in on the talk, stood up and declared that Hawking had omitted a term. Gell-Mann was also working from memory.

Despite his disabilities, by the early 1970s Hawking was beginning to travel extensively. His status as a physicist had grown with his work in collaboration with Penrose, and he was frequently invited to deliver talks and address seminars around the world. At the same time as his scientific reputation was building, Hawking's image as a determined fighter, who would go to any extreme to be treated as a normal human being, was spreading far beyond Cambridge.

One of his oldest and closest friends, David Schramm, who is now at the University of Chicago, has a wealth of anecdotes about Stephen's exploits. His favourite recollection from the early seventies concerns the occasion when he first became aware of Stephen's huge potential for enjoying himself. After a conference in New York, Schramm took the Hawkings to a party thrown by a friend in Greenwich Village. Stephen really enjoyed himself, dancing with Jane, spinning his wheelchair around the room and generally having a great time.

Schramm is also happy to dub his friend an incorrigible flirt and to describe his eyes as tremendously expressive. Women, Schramm claims, were always very interested in Stephen long before his international fame brought him wide attention. Indeed, David Schramm's wife, Judy, was tremendously taken by him when they first met and found his ability to convey his personality by facial expression extremely attractive.

Hawking's interest in dancing has never diminished and the annual college parties at Caius would not be the same without his joining in with the other fellows and their partners on the dance-floor. Nowadays, in his elevated position as professor and head of the DAMTP, he is still to be seen at Christmas discos organized by the students, dancing the night away. His energy, both at work and at play, has become a legend. As David Schramm says, Stephen is a real party animal.

Between trips abroad and working on black holes with Roger Penrose, Hawking was collaborating with George Ellis on a book eventually to be called *The Large Scale Structure of Spacetime*. The idea for the book had arisen back in 1965, when Hawking was still working towards the completion of his PhD. Ellis remembers that the two of them had drawn up a list of future plans, which included 'getting married' and 'writing a cosmology book together'. Because both of them were busy with other projects and domestic changes, work on the manuscript went very slowly. Ellis spent some time in Hamburg and then in Boston, and the two of them began to see each other less frequently. Through Dennis Sciama they managed to secure a contract with Cambridge University Press, which was just starting a series of high-level research monographs aimed at professional physicists.

It took six years to finish the manuscript. They divided up the various topics between them and worked independently, meeting when they could to go through each other's contributions and make changes where appropriate. Ellis did all the typing; when Hawking could no longer write, he dictated his material to Ellis who wrote it up for him. George Ellis was one of Hawking's close associates who could understand his speech, but even he found it difficult at times. He soon discovered that it was much easier to follow what Hawking was saying in discussions about scientific matters, when the conversation consisted largely of familiar technical terms. It was in everyday conversations, which could be about almost anything, that the going got tough.

Because *The Large Scale Structure of Spacetime* took so long to write, events overtook it in a number of areas. In particular, Hawking's own work on black holes (with which Ellis was not directly involved) had progressed faster than they could amend the text. The book dealt purely with classical theories of cosmology, but by the time of its publication in 1973, Hawking had made great strides in the quantum interpretation of black hole physics, and it was not until it went into a second edition that they were able to update the text. The book caused quite a stir in academic circles and did a great deal for the general prestige of the series. Indeed, Hawking is now considered by Cambridge University Press to be the most distinguished author in its catalogue.

The book is incredibly complex, completely unreadable except by experts working in the field of cosmology. Hawking and Ellis had no intention of writing a popular book, and their manuscript fitted the requirements perfectly. However, a favourite story in the Science Department at Cambridge University Press recounts an occasion when an associate of Hawking's ventured his opinion of this first publication. Hawking and Simon Mitton were returning to Cambridge from a meeting at the Royal Astronomical Society in London, and happened to be sharing a railway carriage with the radio astronomer John Shakeshaft. As they pulled out of the station, Shakeshaft, who was sitting in the seat opposite Hawking, leaned forward and said, 'Well, I got a copy of your book, Steve.'

'Oh, did you enjoy it?' asked Hawking.

'Well,' Shakeshaft replied, 'I thought I might make it to page 10, but I only got as far as page 4, and I've given up, I'm afraid!'

Despite the complexity of the book, the latest sales figures show that, since its publication, it has notched up 3,500 copies in hardback and over 13,000 in paperback – one of the best selling research monographs ever published by Cambridge University Press.

Simon Mitton, who left the Institute of Astronomy in 1977, is now the Science Director at Cambridge University Press. He has suggested that the book has sold to a large number of undergraduates who have bought it because it looks good on their book-shelves but have probably never got beyond the second page of tightly packed equations. *The Large Scale Structure of Spacetime* and other, later, technical books of Hawking's showed a definite upturn in their sales curves upon the publication, many years later, of *A Brief History of Time*. After that the original co-author's name, 'S.W. Hawking', printed on the jacket, was hurriedly changed to 'Stephen Hawking', and the sales figures took another climb.

In the world of black hole research, work was moving forward at a startling pace, and Hawking was in the vanguard. It was becoming more and more clear to him that the purely classical interpretation of black holes was deficient. In September 1973 he visited Moscow. The head of the Institute for Physical Problems of the USSR Academy of Sciences in Moscow was a fiery little man with a bald head and boundless energy called Yakov Boris Zel'dovich. He and his team had been working on black holes, in particular on the way in which they interacted with light. Hawking returned to Cambridge convinced that they were on to something but were going about things the wrong way. As he said many years later, 'I didn't like the way they derived their result, so I set out to do it properly.'[7]

What he then decided to attempt was quite revolutionary. As we saw in Chapter 2, the two great pillars of twentieth-century physics are quantum mechanics and relativity, but they are at opposite ends of the spectrum as far as physics is concerned. They speak a different language, and nobody had managed to reconcile the two theories. But this was exactly what Hawking had set his sights on. It seemed to be the only way forward if he were to explain the behaviour of black holes thrown up by the contradictory ideas of Bekenstein on the one hand and of him and Penrose on the other.

Sorting out the problem was easier said than done. Working

on the equations in his head was difficult enough, but after months of intense work Hawking kept coming up with completely nonsensical results. According to the equations, black holes appeared to be emitting radiation. He, and everyone else at the time, believed this to be impossible. He was still convinced that he was really on to something but took the conscious decision not to discuss the problem with anyone until he had settled the matter one way or another.

Christmas 1973 came and he was still in as much of a mess with the mathematics as he had ever been. He decided to rework the equations. He knew that he had cut corners with some of the derivations and believed that these short cuts may have held the key to the problem. During the Christmas vacation he spent lonely weeks running and rerunning the equations through his mind, forcing himself to use ever more complex processes to eradicate the annoying anomalies. Finally, in January 1974, he took the plunge and confided in Dennis Sciama who was organizing a conference at the time. To Hawking's surprise, Sciama was very excited by the idea and, with Hawking's permission, set about spreading the word.

A few days later it was Hawking's thirty-second birthday, and his family arranged a dinner-party to celebrate. Soon after the meal was served, the phone rang. It was Roger Penrose calling from London – he had heard the story propagated by Sciama and wanted to know all about it. The discussion went on and on. The food grew cold, and the other guests waited patiently for Hawking to return to the table. Forty-five minutes later, with the meal ruined, he hung up. Penrose was tremendously excited and wanted to discuss it further.

Going against all current ideas about black holes, by the power of mathematical reasoning Hawking had been forced to the unarguable conclusion that not only did tiny black holes emit radiation but under certain conditions they could actually explode.

By late January one of his colleagues and friends from postgraduate days, Martin Rees, was convinced that Hawking had made a great discovery. Inspired by his latest discussion with Stephen, he bumped into Dennis Sciama in a corridor at the Institute of Astronomy. 'Have you heard?' he said, excitedly, 'Stephen's changed everything!'

Sciama dashed off to see Hawking. By the end of the conversation, he too was convinced and persuaded his former student to announce his results at the conference he was organizing in February at the Rutherford–Appleton Laboratory outside Oxford.

Hawking was driven to the Laboratory through the icy chill of midwinter Oxfordshire and assisted into the building by one of his research students. Sitting patiently to the side of the main group, he listened to the other speakers announcing their latest news. As usual, he asked his customary penetrating questions, trying hard to control his great feeling of excitement. He had a hunch, now supported by a number of his respected colleagues and peers, that he was on to something very big. At last he was wheeled to the front of the lecture theatre and his illustrations were projected on to the back wall while he delivered his talk in the almost unintelligible tones to which his colleagues had become accustomed. His final line was delivered. A stunned hush fell over the entire room. You could hear a pin drop as the audience of scientists tried to absorb the astonishing news. Then the backlash began.

The moderator of the meeting, the English theorist John G. Taylor, jumped up from his seat and proclaimed that what Hawking had said was complete nonsense. Pausing only to drag one of his colleagues from the seat beside him, Taylor stormed from the room and immediately started writing a paper denouncing Hawking's claim. Hawking had expected a reaction but nothing like this. He simply sat on the podium in shocked silence.

John Taylor's paper was dashed off and sent to the scientific journal *Nature* for publication. The editor of *Nature* sent the draft manuscript to Hawking for his comments before making the decision to publish it. Hawking wrote back to recommend publication. He would not want to stand in the way of anyone rash enough to disclaim his work without having investigated the matter thoroughly.

A month after the meeting outside Oxford, Hawking published in *Nature* his own paper describing the newly discovered phenomena. Within weeks, physicists all over the world were discussing his work, and it became the hot topic of conversation in every physics laboratory from Sydney to South Carolina. Some physicists went so far as to say that the new findings constituted the most significant development in theoretical physics for years. Dennis Sciama described Hawking's paper as 'one of the most beautiful in the history of physics'. The radiation that he had discovered could be emitted by certain black holes was from then on known as Hawking Radiation.

However, not everyone was convinced, and it was quite a while before many groups working around the world came to terms with this revolution in black hole physics. It took until 1976 for Zel'dovich's team in Moscow to accept the new ideas. Zel'dovich ran his Institute in an extremely dictatorial manner. What he said went. When he finally gave his endorsement to the theory, his team were compelled to go along with it, just as they had followed when he had disagreed with it.

At the time of Zel'dovich's change of heart, Roger Penrose was invited to Moscow to give a talk that Zel'dovich, as Penrose's colleague and head of the Institute, would be attending. In his lecture notes Penrose had assumed the validity of Hawking's deductions and had built his talk around them. When he arrived, a day before the lecture, he was told bluntly that Zel'dovich did not agree with Hawking; nor did any of his students. Not only that, but he would prefer it if Penrose did not mention Hawking's findings. Penrose was completely thrown. It meant, quite simply, that he had to rewrite his lecture; he set to work, labouring into the small hours. Then, a few hours before he was due on the podium, an assistant turned up at his hotel to inform him that Zel'dovich had changed his mind about Hawking – and so had all his students.

Another story relates how the American physicist Kip Thorne was in Zel'dovich's apartment when the transformation in his thinking actually occurred. Zel'dovich was pacing the room when Thorne arrived, and in a theatrical display of resignation the Russian physicist threw up his arms in despair and said, 'I give up, I give up. I didn't believe it, but now I do.'[8]

The mid-seventies saw the beginnings of a renaissance in public awareness of science, and the idea of such exotic objects as black holes which could eat whole solar systems for breakfast caught the public imagination. It was at about this time that the name of Stephen Hawking first impinged on popular awareness. It was also the time when a great deal of hot air was circulated around the serious theories by writers over-popularizing the ideas the physicists were propounding.

Hawking himself began to pass as a metaphor for his own work. He was becoming the black hole cosmonaut trapped in a crippled body, piercing the mysteries of the Universe with the mind of a latter-day Einstein, going where even angels feared to tread. With the arrival of black holes in the public consciousness, the mystique which had begun to gather around him in Cambridge at the end of the sixties started to extend beyond the cloistered limits of the physics community. Newspaper articles and TV documentaries about black holes started to appear, and Stephen Hawking began to be seen as the man to talk to.

It was not only the media that were beginning to register

what was going on. Hawking's achievements had been noticed by the scientific establishment. In March 1974, within weeks of the announcement of Hawking Radiation, he received one of the greatest honours in any scientist's career. At the tender age of thirty-two, he was invited to become a Fellow of The Royal Society, one of the youngest scientists in the Society's long history to be given such an honour.

The investiture took place at the headquarters of The Royal Society, at 6 Carlton House Terrace, a white-colonnaded mansion overlooking St James's Park in the West End of London. It is traditional for new Fellows of the Society to walk to the podium in the large meeting-room which dominates the building in order to sign the roll of honour and shake the President's hand. However, in Hawking's case, the President at the time, the Nobel Prize-winning biophysicist Sir Alan Hodgkin, brought the membership book down to the front row for him to sign. It took an age for Hawking to sign his own name. The letters were slowly and agonizingly formed on the page alongside the others invested at the same ceremony. As he wrote, the room was completely silent. Then, as he finished the last letter and Hodgkin lifted the book from his lap, the gathered scientists burst into thunderous applause.

The local newspaper, the *Cambridge Evening News*, reported the great event on the day of Hawking's investiture, and a party was thrown at the DAMTP after the ceremony in London. Friends, family and colleagues in the department were all invited to celebrate his achievement. As one of the senior members of the gathering and Hawking's old supervisor, Dennis Sciama was invited to give an impromptu toast to his most successful student, in which he paid tribute to Hawking's achievements and raised his glass to future successes.

As his friends and family joined Sciama in the toast, Hawking surveyed the room. He had come a long way, he knew that, but this was just the beginning. Although he would always believe his investiture into The Royal Society to be the proudest moment of his career, there were plenty more rungs to climb on the career ladder. And, despite the adversities – or perhaps, as some have suggested, because of them – he would continue to climb. Where his feet could not go, his mind would soar.

9

When Black Holes Explode

In 1970, as we have mentioned in Chapter 7, Hawking had shifted the focus of his scientific attention from what goes on at the heart of a black hole, at the singularity, to events that occur on the horizon surrounding the black hole, the nearest thing it has to a 'surface'. A key difference between these studies and the investigation of singularities is that, whatever your theory predicts about things going on at a singularity, you can never test the theory by looking at a singularity because they are all hidden inside black holes (except, of course, the Big Bang singularity at the beginning of time, which Hawking was to investigate more fully later in his career). But when you apply your theory to predict what goes on at the surface of a black hole, at the horizon, then whatever strange events it describes ought to make their mark on the outside Universe, and might even produce effects that could be detected by instruments here on Earth, or on satellites in orbit around the Earth.

It was, in fact, satellite-borne instruments that identified, at about this time, the first really plausible black hole candidate in our Milky Way Galaxy. Just as great new discoveries in astronomy had come about in the 1960s through the investigation of the radio part of the spectrum, at wavelengths longer than those of light, so great new advances came in the 1970s through the investigation of the X-ray part of the spectrum, at wavelengths much shorter than those of light. Unlike radio waves, however, X-rays from space are blocked by the Earth's atmosphere, and do not reach the ground (which is just as well, or we would all be fried). So X-ray astronomy came of age as a branch of science only when suitable detectors were placed in orbit around the Earth. These unmanned satellites transformed astronomers' view of the Universe, showing it to be a much more violent and energetic place than they had thought. And at least some of that violence, they are now convinced, is associated with black holes.

It happens like this. An isolated black hole is, of course, undetectable, except by its gravitational pull – the way it distorts space in its vicinity. It is, after all, black. But a black hole in a binary system, orbiting around a more ordinary star, could make its presence highly visible. Matter torn off the companion star by the gravitational influence of the black hole would funnel down into the hole and be swallowed up. On the way in, it would form a swirling accretion disc, like water going down the plughole of a bath, with gas piling up and getting hot as gravitational energy is converted into energy of motion. It would, calculations showed, get hot enough to emit X-rays.

But how likely is it that a black hole will just happen to be orbiting a companion star? In fact, binary star systems are very common – most stars probably have at least one close stellar companion, and in this our Sun is an exception to the rule. Binaries are also easy to identify because the tug of the two stars on each other makes them wiggle about, producing regular changes which can be observed using telescopes on Earth. The orbital variations also give astronomers a clue to the masses of the two stars, and that turned out to be crucial in identifying black hole candidates.

The snag, for seekers of black holes, is that it is not enough just to identify an X-ray source in a binary system. Both white dwarfs and neutron stars are also compact enough, with a strong enough gravitational pull, to strip matter from a companion and pull it on to themselves, creating hot spots that radiate at X-ray wavelengths.

Several of the first binary X-ray sources found could indeed be identified as white dwarfs, because the orbital variations showed that their masses must be comfortably less than 1.5 solar masses. But four reasonable black hole prospects did emerge from the first X-ray surveys of the sky, carried out in the early 1970s. A first examination showed that all were X-ray sources in binary systems – small, energetic, compact objects orbiting normal stars. More detailed investigations gradually eliminated three of the candidates. One had a mass 2.5 times that of the Sun, and might very well be a neutron star. Another had a mass three times that of the Sun, which seemed a little high for a neutron star, but left room for doubt about its black hole status. The third had a mass only twice that of the Sun. But the fourth had a mass estimated at between 8 and 10 solar masses.

The source is called Cygnus X-1. Only the most tortuous explanations could be invoked to avoid the inference that it harboured a black hole. For example, some astronomers suggested that the unseen companion in the binary system might actually consist of *two* stars – a faint, unseen, ordinary star (too dim to be visible) with a mass six times that of the Sun, itself orbited by a neutron star of 2 solar masses. But the contrived explanations flew in the face of the attractive argument that the simplest explanation was probably the best. Ultimate proof that Cygnus X-1 harbours a black hole would only come if we were able to go and look at it close up; but the weight of evidence that has accumulated over two decades has convinced most astronomers, and the consensus today is that there is a 95 per cent chance that Cygnus X-1 is the first black hole to be identified. Several other promising candidates are also now known, which strengthens the case – we would hardly expect there to be just one detectable black hole in our Galaxy.

The identification of Cygnus X-1 itself as a black hole candidate was the occasion of a famous bet, which sheds intriguing light on Hawking's character. Hawking, whose career has been founded on the study of black holes, made a bet with Kip Thorne, of Caltech (the California Institute of Technology), that Cygnus X-1 does *not* contain a black hole. The form of the bet was that if it were ever proved that the source *is* a black hole, Hawking would give Thorne a year's subscription to *Penthouse*. But if it were ever proved that Cygnus X-1 is *not* a black hole, Thorne would give Hawking a four-year subscription to the satirical magazine *Private Eye*. In June 1990 Hawking decided that the evidence was now overwhelming, and paid up – although, being Hawking, he did so in a typically mischievous fashion, enlisting the aid of a colleague to break into Thorne's office at Caltech. They extracted the document recording the bet, and officially 'signed' his admission of defeat with a thumbprint before returning the paper to the files for Thorne to discover later. Over the following months, Thorne duly received the promised issues of *Penthouse*.

The disparity between the subscriptions wagered simply reflected the different cover prices of the two magazines. But why did Hawking bet *against* black holes? He called it an insurance policy. If black holes didn't exist, he had been wasting his time for most of his career, but at least he would have had the consolation of winning the bet. On the other hand, the only way he could have lost the bet would be if he were right about black holes, so he was happy to offer Thorne some consolation.

In the eyes of most astronomers, Hawking erred on the side of extreme caution in waiting so long to pay up; he had lost his bet, they reckoned, several years ago, for there is no reasonable doubt that Cygnus X-1 is indeed a black hole. And since black holes do exist, that makes Hawking's investigation of their properties during the early 1970s one of the most important pieces of scientific research ever carried out. This work succeeded not only in partially uniting the general theory of relativity and the quantum theory, but also in bringing into the fold the great development of nineteenth-century science, thermodynamics.

Just as Hawking and Penrose had shown that the physics of the Big Bang actually gets *simpler*, not harder, the closer you delve back towards the beginning, so in the late 1960s other research had shown that collapsing black holes are much simpler than the objects that collapse to form them. You could, in principle, make a black hole out of anything: by squeezing the Earth to the size of a pea; or adding scrap iron to a heap until you had enough for gravity to take over; or by watching a star much heavier than our Sun run through its life cycle, explode, and die. But however you make a black hole, what you end up with is a singularity surrounded by a perfectly spherical horizon, with a size (surface area) that depends only on the mass of the hole, not on what it is made of.

This basic truth about black holes was established in 1967, by the Canadian-born researcher Werner Israel. When he first developed the equations, Israel himself thought that because black holes had to be spherical, what the equations were telling him was that only a perfectly spherical object could collapse to form a black hole. But Roger Penrose and John Wheeler found that an object collapsing to form a black hole would radiate away energy in the form of gravitational waves – ripples in the fabric of spacetime itself. The more irregular the shape of the object, the more rapidly it would radiate energy, and the effect of this radiation would be to smooth out the irregularities. Penrose and Wheeler showed that any collapsing object would end up perfectly spherical by the time it formed a black hole. The only thing that could affect the appearance of the horizon surrounding the hole, apart from the amount of matter inside it, is rotation. A non-rotating hole is perfectly spherical, while a rotating hole bulges at the equator.

So, it was established by the early 1970s that a black hole could rotate, but it could not pulsate (Hawking played a small part in this work, too). The size and shape of a black hole depend only on its mass and the speed at which it rotates; the horizon, all that we can see from the outside Universe, carries no identifying features that can tell us what the hole was made of. This lack of identifying features is called the 'no hair' theorem by physicists. A black hole has no 'hair' in the sense that it has no identifying features, and because all we can ever know about it is its mass and its rate of rotation, this makes the mathematical study of black holes much simpler than scientists had feared it would be.

As nothing can get out of a black hole, its mass can never decrease. So the discovery that the surface area of the horizon can never decrease may not seem that dramatic to ordinary mortals. But Stephen Hawking tells how the moment this hit him was so dramatic that it has stuck in his memory for more than twenty years. It happened, as we mentioned in the last chapter, one evening in November 1970, not long after the birth of his daughter Lucy, as he was getting ready for bed. The idea was so exciting that he spent most of the night thinking about the implications.

He was so excited largely because he and Penrose had only just, at that time, come up with a practical mathematical definition of a black hole horizon in terms of the tracks of light rays through spacetime. With this definition, he realized, the surface area of the black hole would always increase if matter or radiation fell into the hole, and even if two black holes collided with one another and merged, the area of the new black hole would always be greater than (or, just possibly, the same as) the areas of the two original black holes put together.

This discovery may have made Hawking so excited that he

could not sleep, and it may have impressed Roger Penrose when Hawking telephoned him the next day to discuss the idea, but initially it made very little impression on other astronomers and physicists, who regarded such notions as rather esoteric. After all, the X-ray observations that led to the identification of Cygnus X-1 with a visible star were made the next year, in 1971, and it was not until the end of 1972 that the consensus that the X-rays come from a black hole orbiting that star was reached. What really began to make other physicists sit up and take notice of Hawking's ideas about the increasing area of a black hole was the seemingly outrageous suggestion that this might be connected with the branch of physics known as thermodynamics.

Thermodynamics is simply the study of heat and motion, as the name implies. It was developed as a branch of science during the nineteenth century, and was of great immediate practical value in the age of steam engines. It rests upon some simple, basic rules, such as the fact that heat cannot flow from a cold object to a hot one (immortalized by the musical duo Flanders and Swann in the memorable couplet 'Heat won't flow from a colder to a hotter/ You can try it if you like but you'd far better notter'). But thermodynamics goes far beyond the day-to-day practicalities of making steam engines work more effectively, and leads on to fundamental truths about the nature of time and the fate of the Universe. One especially important concept, closely linked to the inability of heat to flow 'from a colder to a hotter', is known as entropy.

In everyday language, entropy is the law that tells us things wear out. Hot things cool off as time passes by, and heat flows out of them. Buildings fall down and crumble away; living things grow old and die. These changes are linked to the passage of time, marking a distinction between the past and the future. They correspond to an increase in the amount of *disorder* in the Universe. This disorder is measured in terms of entropy. The flow of time from the past to the future means that the entropy of the Universe must always increase. The same applies to any closed system – the amount of entropy can only increase (or, at best, stay the same); it can never decrease. Now, obviously, the presence of living things on Earth runs counter to this rule. We create order out of disorder by building houses, and so on. But the point is that the Earth is not a closed system. It 'feeds' off the energy flowing from the Sun, dumping entropy as a result. If you take the whole Solar System and treat it as a closed system, the entropy does increase, just as the laws of thermodynamics require.

So Hawking's dramatic realization, coming with such force that evening in November 1970, was to lead to the idea that the law which says that the area of a black hole can only stay the same or increase is equivalent to the law which says that the entropy of a closed system can only stay the same or increase. But even Hawking didn't make that connection at first.

This is the kind of step that is quite often made in science by a junior researcher, not yet hidebound by tradition. The thought of trying to make a connection between the gravitational physics of black holes and the thermodynamic physics of Victorian steam engines would have daunted even the genius of a Hawking. But to a research student, just setting out on a scientific career and faced with two pieces of information that seem to say the same kind of thing in different ways, the similarity seemed worth remarking on.

Of course, research students very often remark on odd similarities and coincidences in science, and most of the time it turns out that there is nothing significant in the 'discovery' at all. But when a student at Princeton University, Jacob Bekenstein, suggested that the size of the horizon around the singularity might literally be a measure of the entropy of a black hole, he started an avalanche of investigation which led Hawking to the discovery that black holes are not necessarily black after all – they explode.

Just as research students are expected to come up with wild ideas (most of which prove fruitless), so it is a common theme in science that some of the most important developments are a result of somebody trying to prove that somebody else's theory is wrong. This happened to good effect in the 1950s and early 1960s, when Fred Hoyle backed a rival model to the Big Bang, the steady state hypothesis, and became its most vocal proponent. Astronomers determined to prove Hoyle wrong worked much harder at establishing the accuracy of the Big Bang model than they might have done had there been no rival on the scene. But sometimes the effort can rebound.

Hawking was annoyed by Bekenstein's suggestion. Even a research student ought to have realized that there is a direct connection between entropy and temperature, so that if the area of a black hole were indeed a measure of entropy it would also be a measure of temperature. And if a black hole had a temperature, then heat would flow out of it, into the cold (−270 °C) of the Universe. It would radiate energy, contradicting the most basic fact known about black holes, that nothing at all – not even electromagnetic radiation – can escape from them. Together with Brandon Carter and Jim Bardeen, Hawking wrote a paper, published in *Communications in Mathematical Physics*, pointing out this seemingly fatal flaw in Bekenstein's suggestion. It gave the formula for working out the temperature of a black hole according to this ridiculous notion, and was published in 1973. But far from agreeing with Bekenstein, the team commented, 'in fact the effective temperature of a black hole is absolute zero . . . no radiation could be emitted from the hole.'[1]

Within a year, however, Hawking had changed his mind. The reasons why he had second thoughts were related to another line of research on black holes he had been pursuing: the possibility, first aired in 1971, that very small 'miniholes', smaller even than the nucleus of an atom, might have been produced in the Big Bang, and could still be at large in the Universe today.

The critical mass needed to make a black hole simply by an object collapsing under its own weight is, as we have mentioned, about three times the mass of the Sun, and the Earth itself would become a black hole if it were squeezed down to about a centimetre. But absolutely anything will make a black hole if it is squeezed hard enough – a bag of sugar, a coin, the book you are reading, *anything*. The difficulty is that, the lighter the object you want to make into a black hole, the harder you would have to squeeze it.

Hawking reasoned that as we look back in time towards the beginning, we look back to higher and higher densities and pressures. So, if we look back far enough, we come to a time when the pressure was great enough to squeeze any amount of matter you fancy, even a few grams, into a black hole.

The one snag with this argument is that if the Universe were perfectly smooth and uniform back then, no miniholes could form; the only black hole would be the entire Universe itself. But, provided there were some irregularities, some variations in density from place to place in the early Universe, then at the appropriate stage of the Big Bang a few grams of matter, any region that just happened to be a little denser than the average, could indeed get pinched off from the rest of spacetime, forming tiny black holes that would last forever (or so Hawking thought in 1971) and still be around today.

We know that the Universe cannot have been perfectly smooth and uniform in the Big Bang, because if it had, there would be no way that irregularities such as galaxies could have formed as the Universe expanded. There must have been 'seeds', in the form of tiny irregularities, on which galaxies could grow by gravitational attraction. So Hawking's notion of primordial black miniholes seemed plausible, even if there was no obvious way to test the idea.

In fact, although lightweight by the standards of conventional black holes, even a minihole may have rather a lot of mass by everyday standards. A black hole weighing about a billion tonnes, for example (the mass of a mountain here on Earth) would have a radius roughly the same as that of a proton. Less massive miniholes would be correspondingly smaller. And if you are dealing with objects as small as that, physicists knew, you have to use the quantum description of reality in order to understand what is going on.

Now, the plot began to thicken. In 1969 Roger Penrose had shown that it is possible for a *rotating* black hole to lose energy, and slow down as it does so. The way this happens is rather like the way in which space scientists sometimes use the gravitational pull of the planets to speed up spacecraft moving around the Solar System. For example, at the time of writing a probe called Galileo has just undergone a 'slingshot' manoeuvre around the Earth, and will eventually, if all goes well, end up in orbit around Jupiter. But in order to get there it will have followed a circuitous route.

After its launch, Galileo was sent not outwards through the Solar System towards Jupiter, but inwards to fly by Venus. By diving around Venus on a carefully calculated orbit, the spacecraft gained energy and speed, and was deflected towards the Earth. Venus lost a corresponding amount of energy, but, being vastly more massive than the space probe, slowed down in its orbit by only a minuscule amount. At the end of 1990, the speeding Galileo carried out another slingshot manoeuvre, this time involving the Earth, and entered an orbit which will bring it back for a second slingshot past the Earth some two years later. Only then will it be moving fast enough to reach Jupiter in a reasonable time – and it is a sign of how much the probe's speed will have been increased that it will reach Jupiter sooner, even after years of delicate manoeuvring to take advantage of the three slingshots, than if it had gone straight out through the Solar System when it was launched.

Penrose showed that similar gravitational effects could boost the energy of electromagnetic radiation near a rotating black hole. The radiation gains energy; the rotation of the hole slows down. In 1973 two Soviet researchers, Yakov Zel'dovich and Alex Starobinsky, extended this idea to show that a rotating black hole should also throw off particles. Their argument had to do with the uncertainty principle of quantum physics, and we shall explain it shortly. They persuaded Hawking that the effect would be real, and he set about trying to find a precise mathematical treatment to describe the phenomenon. He was surprised, and at first annoyed, to discover that the equations said that the same process should be at work even for a non-rotating black hole.

'I was afraid,' Hawking wrote in *A Brief History of Time*, 'that if Bekenstein found out about it, he would use it as a further argument to support his ideas about the entropy of black holes, which I still did not like.'[2] In 1977, he wrote in the January issue of *Scientific American* that he 'put quite a lot of effort into trying to get rid of this embarrassing effect',[3] but to no avail. In the end, Hawking had to accept the mathematical evidence rather than his prejudices. He had found that all black holes emit energetic particles, and that therefore every black hole has a temperature. The temperature exactly matches the thermodynamic predictions related to the surface area of the black hole. We shall now describe how it works (leaving out the detailed mathematics).

Quantum uncertainty doesn't just mean that human instruments are incapable of measuring any quantity precisely. It means that the Universe itself does not 'know' any quantity with absolute precision. This applies to energy, as much as to anything else. Although we are used to thinking of empty space as containing nothing at all, and therefore having zero energy, the quantum rules say that there is some uncertainty about this. *Perhaps* each tiny bit of the vacuum actually contains rather a lot of energy.

If the vacuum contained enough energy, it could convert this into particles, in line with $E = mc^2$. But things are not as simple as this. If the hypothetical energy of uncertainty in the vacuum were converted into particles, and the particles became permanent features of the Universe, then the rules of uncertainty would be violated – both human observers and the Universe would now be certain that there was something, in the form of a particle or two, where previously there had been nothing. Uncertainty works two ways: it is just as forbidden to be certain that the energy is non-zero, in these circumstances, as it is to be certain that the energy is zero.

In fact, the precise version of the uncertainty rule says that energy can only be 'borrowed' from the vacuum for a very short time, a time determined by Planck's constant. This is related to the uncertainty inherent in the measurement of time itself. The only way in which this energy can then be converted into particles is if particles are always created in pairs, which then interact with one another and annihilate themselves before the Universe has time to 'notice' that the energy has been borrowed. This means that the particles created out of the vacuum are matched in a special way.

Every variety of particle, such as an electron, has a counterpart known as an antiparticle (in the electron's case, a positron). Antiparticles have been manufactured in experiments using particle accelerators, and they are also found in cosmic rays (energetic particles reaching the Earth from space), as well as being predicted by quantum theory, so there is no doubt that they exist. In many ways, an antiparticle is a mirror image of its particle equivalent: the positron, for example, carries positive charge, whereas the electron carries negative charge. And whenever a particle meets its antiparticle counterpart, the two annihilate each other.

So, according to quantum theory the vacuum is a seething sea of 'virtual' particles. Pairs such as electron–positron are constantly being created, interacting with one another and disappearing in accordance with the quantum rules. Overall, no energy is released, but virtual pairs flicker in and out of existence all the time, below the threshold of reality.

What Hawking showed was that, even for a non-rotating black hole, this process can drain off energy from a hole and release it into the Universe at large. What happens is that a pair of virtual particles is created just outside the horizon of the hole. In the tiny fraction of a second allowed by quantum uncertainty, one of the particles is captured by the hole. So the other particle has nothing to annihilate with, and escapes, carrying energy with it.

Where has the energy come from? In effect, it is the gravitational energy of the hole. The energy of the hole creates two particles, but it captures only one of them, so only half the energy debt is repaid and the net effect is that the hole loses mass. Other things being equal – if the hole does not gain mass from somewhere else – it will steadily shrink away as a result, evaporating like a puddle in the sunshine. This process is slow but sure, taking billions of years to shrivel even a proton-sized minihole to the point where it explodes. Hawking had contradicted his own earlier conclusion that the surface area of a black hole cannot decrease. Having established a link between black holes and thermodynamics by showing that, according to general relativity alone, black holes cannot shrink, he had now found that if you add quantum theory to the brew the link with thermodynamics is strengthened, but now black holes *must* shrink.

For ordinary black holes, made out of dead stars, this effect would be of no real importance. A black hole with three or four times the mass of our Sun and a horizon roughly as big as the surface of a neutron star will be constantly swallowing traces of gas and dust from its surroundings, even in the depths of space, and it is simple to show that the mass lost by Hawking Radiation is much less than the mass gained by this accretion. If nobody had thought of the notion of miniholes, nobody would have been very interested in Hawking Radiation. But since Hawking had already come up with the notion of miniholes, the idea of quantum evaporation of black holes made an immediate impact.

A hole smaller than a proton will not eat up much material from its surroundings, even if it happens to be inside a planet. To a hole that small, even solid matter is mostly empty space! So the Hawking Radiation from the surface of a minihole will actually dominate its behaviour. Hawking showed that the radiation produced in this way gives the hole a temperature, exactly the temperature suggested by the work of Bekenstein. For a black hole with the mass of our Sun, this temperature is about one ten-millionth of a degree K (with the resulting ultra-feeble Hawking Radiation easily overwhelmed by in-falling matter); but for a minihole with a mass of a billion tonnes and the size of a proton, the temperature is about 120 billion K. As these examples indicate, the temperature depends on one over the mass of the hole, so as it loses mass and gets smaller such a hole gets hotter and radiates energy faster, until it finally explodes in a burst of X-rays and gamma-rays.

Science-fiction fans may be intrigued to know that if we could find a proton-sized minihole today, it would be a more than useful energy source. The output from such a hole would be about 6,000 megawatts, and could make a substantial contribution to the energy requirements of even a large country. Unfortunately, though, holding on to such a hole if you found it would be tricky – remember that it would weigh a billion tonnes, and that gravity would tend to pull it down towards the centre of the Earth.

The lifetime of such a minihole depends on the exact mass it starts out with, but roughly proton-sized black holes born in the Big Bang should be exploding here and there in the Universe today. Intriguingly, detectors flown on satellites have reported occasional bursts of gamma-radiation coming from the depths of space, and there is no universally accepted explanation for this phenomenon. It is just possible that the Hawking Radiation from exploding black holes has actually been discovered, although it will be almost impossible ever to prove this.

Hawking had achieved something that even he had thought to be almost impossible, using a combination of general relativity and quantum physics (plus a smattering of thermodynamics) in one package to describe a physical phenomenon. It was this work that made his name outside the specialist circles of mathematicians and astronomers, and any physicist today can tell you what Hawking Radiation is, and why it is important. But in a quirky gesture which is in some ways typical of Hawking's attitude towards established conventions, the astonishing discovery that 'black holes are not black' was announced first not in the pages of a scientific journal such as *Nature*, but in an essay that Hawking entered for a somewhat obscure competition organized by the Gravity Research Foundation in America.

The Gravity Research Foundation runs an annual competition for articles describing new research into the nature of gravity. Until the 1970s, it had been almost exclusively a domestic US competition, with very few entries from abroad, although it had once been won by an expatriate Briton living in the USA. Then, with his last contribution to academia, one of us (J.G.) won the prize in 1970. So when Stephen Hawking won the same prize a year or two later for an essay describing black holes, J.G. quickly sent him a congratulatory note. It was nice, the note said, to see Hawking's name on the list of prizewinners, because this added to the prestige of the award, and gave previous winners a chance to bask in the reflected glory. 'I don't know about the prestige,' Hawking wrote in reply, 'but the money's very welcome.'

The 'official' version of the exploding black hole story appeared first in *Nature* on 1 March 1974.[4] While the Gravity Research Foundation essay carried the dogmatic title 'Black Holes Aren't Black', the *Nature* paper, uncharacteristically for Hawking, was equivocally headed 'Black Hole Explosions?'. It sparked a furious debate, as we saw in Chapter 8, with some opponents of the idea suggesting that this time Hawking really was talking rubbish. John Taylor and Paul Davies, of King's College in London, combined to produce a retort in the issue of *Nature* dated 5 July 1974,[5] headed 'Do Black Holes Really Explode?' and answered their own question with an unequivocal 'No'. Even Taylor and Davies, though, were soon persuaded that they were wrong and Hawking was right.

More important even than the specific idea that black holes explode was the underlying basis for this discovery, that quantum physics and relativity could be fruitfully combined to give us new insights into the workings of the Universe. Soon, Hawking would be using that insight to focus, once more, on the puzzle of the singularity at the beginning of time. But it seems, with hindsight, singularly appropriate that his election as a Fellow of The Royal Society, Britain's highest academic honour, should have come in the spring of 1974, within a few weeks of the publication of the *Nature* version of the exploding black hole paper. Ten years after being given just two years to live, however (and scarcely five years after the deterioration that had seemed likely to cut short his promising career), Hawking's research was really getting into its stride. In the second half of the 1970s he moved on to investigate the origin of the Universe itself, going back to the beginning of time.

10

The Foothills of Fame

Reflecting on his achievements during the first thirty-two years of his life, Stephen Hawking must have felt a deep sense of pride in what he had accomplished. The 1970s were the years when he established himself as a world-class physicist, and they marked the beginning of two decades of startling success in the disparate worlds of arcane research and popular writing.

Soon after becoming a Fellow of The Royal Society, Hawking was invited to spend a year away from Cambridge at Caltech, in Pasadena. The research year, funded by a Sherman Fairchild Distinguished Scholarship, was to study cosmology with the eminent American theoretician Kip Thorne.

Pasadena is a leafy suburb of Los Angeles, nestling up against the San Gabriel mountains to the north-east of Hollywood. The wide boulevards intersecting the district are lined with grand old houses, and in the heyday of Hollywood it was a favourite haunt of film-stars. The main street, Colorado Boulevard, was immortalized in the Jan and Dean song 'Little Old Lady from Pasadena', and there has been no shortage of celebrity names who have taken up residence there over the decades. However, in the summer Pasadena is one of the smoggiest areas of Los Angeles because the escape of ozone is inhibited by the mountains. If a Stage 2 Smog Alert is sounded, citizens are advised to stay indoors unless on essential business, and the authorities have the power to make industry and commerce temporarily shut down. Smog alert warnings are broadcast on the radio, and illuminated signs are switched on over freeways. Perhaps the American Indians displayed great powers of premonition, when, long before white men arrived, they named the region 'Valley of the Smokes'.

Caltech itself is unique in that, for such a prestigious institution, it is tiny. In the mid-seventies it was home to no more than fifteen hundred students and was a tenth the size of colleges with comparable reputations such as Harvard or Yale. But despite its size, Caltech is the West Coast's Mecca for science and technology. Throughout its history it has attracted the leading people in their fields from all over the world. Nobel Prize-winning physicist Robert Millikan arrived there in the twenties, and was frequently visited by Albert Einstein. Money simply pours into the place from benefactors ranging from private individuals fascinated with scientific research to multinationals such as IBM and Wang. With some of the best telescopes in the world a matter of miles away on Mount Wilson and the massive Jet Propulsion Laboratory as a gargantuan 'annex', dwarfing the mother campus, it has everything a scientist could wish for.

Some of the world's best physicists were based at Caltech in the seventies. Kip Thorne headed the relativity group there, and the charismatic Nobel laureate Richard Feynman still taught there, and played bongos in college bands during the evenings. Academic quality aside, the contrast between Caltech and Caius could not have been starker. The buildings making up the campus, although tastefully designed and constructed in sand-coloured stone, are all Spanish-style, light and airy, with the nine-storey Millikan Library block rising at the centre. Those admitted to Caltech are among the very best students in the country, and they are driven hard. There is very little social life on campus, and the suicide rate among students ranks almost as high as its academic reputation.

Having said that, there was no shortage of colourful characters around the place at the time of Hawking's sabbatical.

Richard Feynman, a physics professor, had already acquired a formidable reputation as an amiable eccentric and once took on the local authorities who were trying to close down a topless bar in Pasadena. In court he claimed that he frequently used the place to work on his physics. Feynman and Hawking shared an offbeat sense of humour, and although their work rarely overlapped they had a lot of time for each other. Both men have achieved international fame as scientists and live-wire characters, and each has acquired cult status in the wider world outside his own discipleship of graduate students and fascinated lay-people. When Feynman died of cancer in 1988, the whole of Caltech mourned and the global village of science felt the loss.

Kip Thorne, now viewed as the West Coast's relativity guru, favours floral shirts, beads and shoulder-length grey hair. He introduced Hawking to another physicist who was to play a significant role in collaborations and become one of Hawking's lifelong friends – Don Page. Page, who was born in Alaska and graduated from a small college in Missouri, was working on his PhD at the time of Hawking's visit. The two of them immediately hit it off, and before Hawking's year at Caltech was over they had written a black hole paper together.

The family were excited by the move. Jane organized all the details, booking airline tickets, packing and arranging schedules, as well as managing to transport a severely disabled husband and two young children to the other side of the world almost single-handedly. At Caltech Hawking was treated with the respect he should have received at his own college in Cambridge. Wooden ramps were fitted against the kerbs in the vicinity of his office so that he could get around easily in his wheelchair, and he was provided with a smart office and every aid and resource he would need to help him with his research. The work was satisfying, and he found collaboration with Thorne's team both stimulating and scientifically rewarding. Jane and the children enjoyed the southern Californian climate. Despite the air pollution, noise and traffic congestion of Los Angeles, the beaches and the blue Pacific made a welcome change from the often monotonous lifestyle and erratic weather of Cambridgeshire.

With her blonde hair, four-year-old Lucy was the epitome of the Californian flower-child, and loved the place. Robert had to continue with his schooling, but there was plenty of time for the family to be together and do at least some of the things they enjoyed back home. Within Caltech's cloistered environment, the family were sheltered from the extremes Los Angeles had to offer and, moving in privileged academic circles, Pasadena was not unlike the cosiness of Cambridge – but with sunshine. Jane took the children to Disneyland, and Stephen joined them to travel around southern California when he could take time off from his research. Friends and colleagues would often visit. They took trips in hired cars to Palm Springs and resorts along the coast, as well as getting to see a little more of America between duties at Pasadena.

Back in Britain, the government had finally agreed to join the European Common Market by the end of the decade and oil had begun to flow from the North Sea rigs. It seemed that the early-seventies gloom of strikes, power-cuts and the three-day week may at last have begun to lift. American astronauts and Soviet cosmonauts shook hands hundreds of miles above a burning Cambodia. Returning to England in 1975, the family were ready for changes and improvements in their own lives.

It often takes a protracted change of lifestyle to highlight the alterations that can be made when things return to the old routine, and the Hawkings saw immediately that they did not want to go back to the old pattern of life in Cambridge.

In some ways they were glad to be back home. The countryside was greener, the weather less predictable, the television less obtrusive and the tea tasted as it had been ordained by God to taste. But the simple fact was that, having experienced the comforts of California, they were no longer prepared to put up with some of the inconveniences of their lives in Cambridge.

The first thing that hit them was that, quaint and nostalgic as it may have been, the house in Little St Mary's Lane was far too small for them. Stephen was finding it impossible to use the stairs, and it was too cramped for a family of four. Hawking asked the College to help them find somewhere more suitable for their needs. On this occasion, the authorities were more than willing to come to their assistance. As Hawking puts it, 'By this time, the College appreciated me rather more, and there was a different Bursar.'[1]

They were offered a ground-floor flat in a large Victorian house owned by the College, in West Road, not far from the gate of King's College and a mere ten minutes' wheelchair ride from the DAMTP. The house has a large garden, regularly tended by college gardeners who kept it in a permanent state of elegance. The children loved it, and there was never a problem about their playing on the lawns, an informal truce with the gardeners having been established. Wide doorways made it easy for Hawking to manoeuvre his wheelchair around the entire flat, and because it was all on one level he no longer had to struggle upstairs to get to the bedroom.

By 1974, Hawking was having difficulty getting in and out of bed and feeding himself. Until their return from the States, Jane had been Stephen's unpaid, twenty-four-hours-a-day nurse, as well as his wife. She had, of course, been fully aware of the responsibilities expected of her when she decided to marry Stephen in 1965, but the effort of bringing up two young children and running the home as well as looking after her husband was beginning to take its toll on her emotional well-being. They decided to invite one of Hawking's research students to live with them in West Road. The flat was big enough for another adult, and in return for free accommodation the student would help Jane to look after Stephen.

The system worked well. In fact, as Hawking's prestige grew it was considered an honour and a good career move to become Hawking's 'student-in-residence'. It was inevitable that close bonds were established between the young research assistant and his mentor. While Jane received much-needed help, the student gained a closer insight into Hawking's mind, and some of his genius was bound to rub off. At least that was the theory. There was, of course, another side to this: as Hawking himself has said, 'It was hard for a student to be in awe of his professor after he has helped him to the bathroom!'[2] Bernard Carr, who was one of Hawking's earliest students to have this honour and is now at the University of London, describes his time there as 'like participating in history'.[3] The duties of the lodgers were manifold. To earn their keep they were expected to play as required the roles of nanny, secretary and handyman, helping with travel arrangements, babysitting the children, drawing up lecture schedules and managing general household repairs.

Another early lodger was the American physicist Don Page. After finishing his PhD at Caltech, Page had written to Hawking asking for a job reference. In the months that followed, several research groups wrote to Hawking about Page, and each time he gave a favourable reference. Then, some time later, he wrote to the young physicist, 'I've been writing letters of reference for you, but I may have a position myself.'[4] Hawking managed to help Page to secure funding for a year, and then organized a grant for a further two years of research. Page joined the Hawking household in 1976 and re-established the close friendship they had enjoyed in California, a friendship which has survived to the present day.

One of Page's duties was to commute with Hawking each day between West Road and the DAMTP. This was seen as a good time to talk, to summarize the previous day's efforts and to consider the tasks for the day ahead. It was a very productive time, even though Page found Hawking's way of working through complex mathematics in his head quite hard to get used to. Talking about the twice-daily journey, he has said:

I found it very good training. During the three years I was a postdoc, I lived with the Hawking family, and a lot of times I'd walk back and forth with him. Of course I couldn't write while I was walking, and sometimes he would ask me something, and I'd try to think it out in my head. When you have to do it in your head, you have to get really to the heart of the matter and try to eliminate the inessential details.[5]

Around the time of the move to West Road, Hawking found that he could no longer use the three-wheel invalid car he had had on loan from the National Health Service since 1969 and in which he travelled to the Institute of Astronomy three times a week. At first this appeared to be another blow, but, as has often been the case with the Hawkings, they were able to turn the situation to their advantage. Jane says:

It was a blessing in disguise, because the roads are so dangerous out to the Institute anyhow. It didn't matter because we could afford to buy the electric wheelchair . . . which he runs along in, and is really much more convenient for him because he doesn't have to be sure of having people to help him in and out as he does with the car. So he's completely independent in the electric wheelchair. There's always some compensating factor that makes deterioration acceptable.[6]

Hawking became a real demon of a wheelchair driver. One journalist described his skills thus:

He hurtles out into the street. At full throttle the chair is capable of a decent trotting pace, and Hawking likes to use full throttle. He also knows no fear. He simply shoots out into the middle of the road on the assumption that any passing cars will stop. His assistants rush nervously out ahead of him to try to minimize the danger.[7]

Jane's relief that he no longer had to use the three-wheeler on the roads of Cambridge could so easily have been misplaced. Indeed, recently, in early 1991, Hawking was involved in an accident in his wheelchair. He is a very familiar figure in the city, and passers-by stop and talk to him. However, on this occasion a driver failed to see the chair with the slumped figure of the world's most famous living scientist at the controls. The car hit the chair, and Hawking's frail body was thrown on to the road. It could have been a disastrous accident, but fortunately he suffered only minor injuries, cutting his face and damaging a shoulder. It is typical of the man that, against medical advice, he was back in his office within forty-eight hours and demanding that his papers and books be propped up in front of him so that he could work.

On other occasions, his 'boy-racer' antics have caused great embarrassment. In June 1989, Hawking was to deliver the prestigious Halley Lecture at Oxford University. A young, newly appointed physics professor, George Efstathiou, was given the unenviable task of looking after the eminent visiting lecturer before, during and after the talk. Hawking arrived at the Department of Zoology, where the University's largest lecture theatre is housed, and was escorted into reception. It was Efstathiou's job to get his famous charge to the theatre, one floor below, where the Vice-Chancellor of the University and six hundred students, city dignitaries and interested lay-people were waiting in expectation.

A two-man lift at the end of the reception area would take them to the floor below and lead, via a short corridor, to the lecture theatre. The lift doors were open. Before Efstathiou had a chance of helping Hawking into the lift, Hawking set the chair to full throttle and headed for the open doors a dozen yards ahead of him.

Efstathiou remembers clearly that he estimated, even from that distance, that Hawking could not make it into the narrow lift entrance, and he could do nothing but watch in horror as his guest speaker hurtled towards the aperture. At last propelled into action, Efstathiou gave chase, but could not catch up. To his amazement Hawking made it through the lift doors.

But that was only the beginning of Efstathiou's troubles. For, as Hawking had entered the lift, the chair had twisted at an angle and jammed in the narrow space. The lift doors closed automatically behind the chair, trapping its wheels between them. Efstathiou was panic-stricken. Downstairs, hundreds of people were waiting for Hawking, who was already late. The disabled scientist could not reach any of the control buttons, but the doors had closed on him. What was to be done?

Meanwhile, seemingly unperturbed by events, Hawking was busily punching instructions into his computer to get it to put the chair into reverse. If Efstathiou could have seen his face, he would undoubtedly have encountered the famous, mischievous Hawking smile. Finally, Efstathiou succeeded in squeezing his arm into the crack between the doors and just managed to reach the door-opening button. Freed, Hawking sent the chair into high-speed reverse and re-emerged unscathed and grinning. As Efstathiou says, 'That experience was quite an initiation into college administration!'

Hawking uses his wheelchair as an appendage to his paralysed body, a device for the physical expression of his personality. He cannot shout and scream at people. Nowadays, of course, his computer-generated voice is totally expressionless, but he can certainly move his wheelchair around. Hawking has, as one journalist has put it, 'a strain of fierceness running through [his] personality, surfacing in spates of impatience or anger'.[8] If he feels that someone is wasting his time, he simply spins his wheelchair on the spot and speeds out of the room in a huff.

John Boslough recalls an incident when he got on the wrong side of Hawking and received the usual rebuff. While talking to him he had become so oblivious to the other's condition that he began talking about a problem he was having with his elbow as a result of a squash match in London the day before. 'Hawking made no comment. He simply steered his wheelchair out of the room and waited in the hall for me to return to the subject at hand – theoretical physics'.[9] Perhaps talking to a paralysed man about squash was not the most subtle of things to do, but the incident illustrates the very well-known fact that Hawking is certainly not a man to cross lightly.

His favourite move, when he is annoyed by something someone has said, is to drive over their toes. By all accounts, a number of his students and colleagues have had to develop pretty fast reflexes. One of Hawking's former students, Nick Warner, claims, 'His great regret is that he's not yet run over Margaret Thatcher!'[10] Perhaps he will get the chance one day.

There is, of course, a very different side to his personality: Hawking the family man. He loves nothing more than using his wheelchair skills when playing with his children, and applies his usual recklessness when racing around the garden of West Road playing tag. The sad fact is that he can play no other physical games with them. It was Jane who taught them cricket and played Stephen's old game of croquet on warm summer evenings with Robert, Lucy and, later, Timothy. As one journalist wrote,

In many ways, she has had to be both mother and father to her children. Even the hours she spent as a schoolgirl on the cricket pitch of St Albans High School, alternately bored to tears and terrified of the ball, were to have their value. 'I have been the one who has to teach my two boys to play cricket – and I can get them out!' she has said.[11]

As their first two children were growing up, Hawking was receiving greater and greater accolades as a scientist. In the space of just two years, 1975 and 1976, he won six major awards. First was the Eddington Medal from the Royal Astronomical Society in London, given the year he returned from California. This was followed shortly by the Pius XI Medal, bestowed by the Pontifical Academy of Science at the Vatican. In 1976 came the Hopkins Prize, the Dannie Heinemann Prize from the USA, the Maxwell Prize and The Royal Society's Hughes Medal – the citation for which noted 'his remarkable results in his work on black holes'. As the international physics community began to recognize his talents, his own university was increasingly acknowledging Hawking's worth. During the move from Little St Mary's Lane to West Road, he was made Reader in Gravitational Physics at the DAMTP, an academic position somewhere between a fellow and a professor.

As the awards and prizes mounted up, Jane was becoming increasingly disillusioned with their life and her role in it. It was a time of great change in the way the West perceived women and their position in society. The sixties, for all their sexual liberation and permissiveness, saw very little real change in the role played by women or the way in which they were treated by the other half of the population. What sexual permissiveness and 'liberation' really meant was a different system by which the average woman could be exploited, the whole thing wrapped in a sugar-coating of freely available contraception and shifted morality.

In the seventies, women gained a little more self-respect. This was in part backed up by changes in the law and the support of the media. Some of these events undoubtedly altered Jane's perception of her role. She was happy to play nurse, support her husband through his glittering career and raise a family almost single-handedly. But she had a growing feeling that she was being ignored as a human being, as an intelligent woman who was academically successful in her own right. She was beginning to feel nothing more than a sidekick to the great Stephen Hawking. As she has put it:

> Cambridge is a jolly difficult place to live if your only identity is as the mother of small children. The pressure is on you to make your own way academically.[12]

Cambridge looks like a quaint little English town, but there is a certain degree of bitchiness within its refined academic élite. Although the university community has always been quick to reinforce the image of Jane Hawking as a caring and devoted mother and wife, an element of professional jealousy does undoubtedly creep in. The claws are only sheathed by a thin veneer of civilization, and while her husband was collecting prize after prize, Jane was sliding into a state of declining self-respect:

> I felt very hurt. I saw myself single-handedly making everything possible for Stephen and bringing up the two children at the same time. And the honours were all going to Stephen.[13]

She decided to do something about it, and embarked on a PhD course in medieval languages, specializing in Spanish and Portuguese poetry. She has said, on reflection:

> It was not a very happy experience. When I was working I thought I should be playing with the children, and when I was playing with the children I thought I should be working.[14]

Jane survived the course and went on to become a schoolteacher in Cambridge. But the feeling, as she puts it, of being 'an appendage' has never left her entirely:

I'm not an appendage, though Stephen knows I very much feel I am when we go to some of these official gatherings. Sometimes I'm not even introduced to people. I come along behind and I don't really know who I'm speaking to.[15]

To be fair to Stephen Hawking, according to his friends and colleagues he has never failed to bolster Jane's contribution to his success and well-being. He takes every opportunity to speak of the great efforts and sacrifices she has made in order to allow them to live as normal a life as possible. One of his great regrets is that he has been unable to play a greater role in helping to raise the children, and he would love to be able to play more than tag and chess with them.

Naturally, Hawking's condition has freed him from many duties other than helping to run the home. His various positions at the University have all come with reduced teaching and administration loads, and he has been allowed to spend a far greater proportion of his time thinking than ever the average professor can manage. Some have attributed his great successes in cosmology to this enhanced cerebral freedom, yet others have claimed that the turning-point in the application of his abilities was the onset of his condition, and that before then he was no more than an averagely bright student. Whatever the reason for his great insight and astonishing grasp of his subject, it may be true to say that he would not have progressed so quickly or soared to such heights if he had been expected to spend vast amounts of time organizing committees, attending faculty meetings and overseeing undergraduate applications.

Feelings of growing resentment over their respective roles within the partnership were not the only difficulties slowly growing into problems for the couple during the seventies. There was the question of religion. Jane was raised as a Christian and has very strong religious views. To one interviewer she has said:

Without my faith in God, I wouldn't have been able to live in this situation. I wouldn't have been able to marry Stephen in the first place, because I wouldn't have had the optimism to carry me through, and I wouldn't be able to carry on with it.[16]

Hawking, for his part, is not an atheist; he simply finds the idea of faith something he cannot absorb into his view of the Universe. His outlook is not unlike that of Einstein, and he has been quoted as saying:

We are such insignificant creatures on a minor planet of a very average star in the outer suburbs of one of a hundred thousand million galaxies. So it is difficult to believe in a God that would care about us or even notice our existence.[17]

It is clear from these two statements alone that the couple have had very different views almost from the moment they met. Jane attributes Hawking's religious views partly to his physical condition:

As one grows older it's easier to take a broader view. I think the whole picture for him is so different from the whole picture for anybody else by virtue of his condition and his circumstances – being an almost totally paralysed genius – that nobody else can understand what his view of God is or what his relationship with God might be.[18]

But is this really the case? There have been many philosophers and scientists throughout history who would have made very similar statements to Hawking's, but they did not suffer from ALS. Equally, of course, there are a number of practising scientists who have very strong Christian convictions, and some have claimed that Hawking is simply not qualified to make statements about religion because he knows nothing about it. But what qualifications does one need? Hawking works in a field which *does* impinge on religion. His work deals with the origins and early life of the Universe. Could a subject be any more religious? He once stated:

It is difficult to discuss the beginning of the Universe without mentioning the concept of God. My work on the origin of the Universe is on the borderline between science and religion, but I try to stay on the scientific side of the border. It is quite possible that God acts in ways that cannot be described by scientific laws. But in that case one would just have to go by personal belief.[19]

And that has never been Hawking's way.

When asked if there is any conflict between religion and science, Hawking tends to fall back on the same argument about personal belief and sees no real conflict. 'If one took that attitude,' he replied, when asked whether he believed that science and religion were competing philosophies, 'then Newton would not have discovered the law of gravity.'[20] And what, in the light of Stephen's and Jane's dilemma, do we make of the famous last paragraph of *A Brief History of Time*?

However, if we do discover a complete theory, it should in time be understandable in broad principle by everyone, not just a few scientists. Then we shall all, philosophers, scientists, and just ordinary people, be able to take part in the discussion of the question of why it is that we and the Universe exist. If we find the answer to that, it would be the ultimate triumph of human reason – for then we would know the mind of God.[21]

Science, it seems, may one day answer the question 'how?' but not 'why?'.

Despite such statements, what really began to cause problems for Jane was a growing feeling that her husband was trying to eradicate any necessity for God in his view of the Universe. And, as his fame and influence grew, she saw this as an escalating problem. It is doubtful that she believed he was fighting any kind of anti-religious crusade with his work or that he was deliberately trying to prove the faithful wrong. It simply seemed to her that, in his Universe, pure mathematical reasoning overrode any need for God:

There's one aspect of his thought that I find increasingly upsetting and difficult to live with. It's the feeling that, because everything is reduced to a rational, mathematical formula, that must be the truth. He is delving into realms that really do matter to thinking people and in a way that can have a very disturbing effect on people – and he's not competent.[22]

But who is? If nothing else, religion is a very personal matter. Are the leaders of the various Churches any more knowledgeable about the origins and meaning of life than a scientist? Why should Stephen Hawking be any less competent to talk about God than the next person – or the next pontiff, come to that? Were the men of God right to sentence Galileo to end his years in solitary misery? Were they right to burn Giordano Bruno at the stake for daring to propose a contrary view of the Universe? Have all the religious wars of human history, with their accompanying terror and misery, been justifiable? Has organized religion been competent in those circumstances?

Jane is not scientifically trained and cannot share her husband's insight into the subject, which he can articulate only with his professional colleagues. She has said:

One of my greatest regrets is that, not being a mathematician, I can understand Stephen's work only in picture terms. He has to keep everything down to earth to explain it to me. It's a good discipline for him.[23]

This had never been a problem before, but when Jane began to see that Hawking was approaching territory whose philosophical foundations were very close to her personal beliefs, it must have set alarm bells ringing.

What she objects to most strongly is Hawking's 'no-boundary' model of the Universe, which suggests that the Universe is self-contained. It is a model with which Hawking is particularly pleased. He has said of the idea, 'It really underlies science because it is really the statement that the laws of science hold everywhere.'[24] When addressing the problem of whether, if the Universe is self-contained, we need to explain how it got there in the first place, his answer is that we do not – 'It would just BE.'[25]

Hawking has at least one close colleague with strong religious convictions, his friend and collaborator Don Page. In fact Page is a born-again Christian, an evangelist as well as a cosmologist. He seems to find no difficulty in marrying the two extreme aspects of his life and work. He says of the no-boundary model:

[In] the Judaeo-Christian view, God creates and sustains the entire Universe rather than just the beginning. Whether or not the Universe has a beginning has no relevance to the question of its creation, just as whether an artist's line has a beginning and an end, or instead forms a circle with no end, has no relevance to the question of its being drawn.[26]

Jane once told a reporter that she had been saddened when, soon after he had taken up residence in their home, Page tried to engage Hawking in a religious discussion but was forced to give up. Despite their vastly differing outlooks, the two men have remained friends, simply agreeing not to discuss any form of personal God.

Hawking confounds both his critics and supporters with seemingly ambiguous statements, such as:

Even if there is only one possible unified theory, it is just a set of rules and equations. What is it that breathes fire into the equations and makes a Universe for them to describe?[27]

Surely Hawking is not here suggesting that there may be a role for a Creator after all. On this matter he seems to take pleasure in leaving things open-ended. By simply limiting the need for a God, he has held back from denying God's existence altogether:

Einstein once asked the question, 'How much choice did God have in constructing the Universe?' If the no-boundary proposal is correct, he had no freedom at all to choose initial conditions. He would, of course, still have had the freedom to choose the laws that the Universe obeyed. This, however, may not really have been all that much of a choice; there may well be only one, or a small number of complete unified theories . . . that are self-consistent and allow the existence of structures as complicated as human beings who can investigate the laws of the Universe and ask about the nature of God.[28]

Thinkers on both sides of the divide – those who support conventional religious views as well as the cynics and atheists – have quoted and misquoted Hawking on so many occasions that one writer recently compared his eloquence and quotability to that of Shakespeare or the Bible. Hawking scoffs at such suggestions, restating the fact that his quotability is derived from his succinctness, a talent he has had to nurture because of the difficulty he has communicating.

Hawking seems to have done little to help Jane through this crisis. She was, and perhaps still is, left exasperated by his stubbornness on the issue. 'I pronounce my view that there are different ways of approaching it [religion], and the mathematical way is only one way,' Jane has said, 'and he just smiles.'[29]

It is not only conventional religion for which Hawking feels extreme scepticism. The lessons he learnt from the ESP experiments in the fifties have never left him, and he has no time for mysticism or metaphysics in any shape or form. A number of writers have made attempts to bridge the gap between mysticism and late-twentieth-century physics. There are many who see parallels between Eastern religion and quantum mechanics, ancient teachings and chaos theories, but Hawking pooh-poohs the whole scene. In his book *Lonely Hearts of the Cosmos*, Dennis Overbye describes an occasion when he met Hawking in the seventies and managed to steer him onto the topic of mysticism without getting his toes crushed. Overbye quoted the anthropologist Joseph Campbell on the Hindu goddess Kali, 'the terrible one of many names whose stomach is a void and so can never be filled, whose womb is giving birth forever to all things'. He then tried to draw a connection between Kali and black holes. Barely able to contain himself, Hawking snorted:

It's fashionable rubbish. People go overboard on Eastern mysticism simply because it's something different that they haven't met before. But, as a natural description of reality, it fails abysmally to produce results ... If you look through Eastern mysticism you can find things that look suggestive of modern physics or cosmology. I don't think they have any significance.

Calling these things black holes was a master-stroke by Wheeler because it does make a [psychological] connection, or conjure up a lot of human neuroses. If the Russian term 'frozen star' had been generally adopted, then this part of Eastern mythology would not at all seem significant. They're named black holes because they relate to human fears of being destroyed or gobbled up. So in that sense there is a connection. I don't have fears of being thrown into them. I understand them. I feel in a sense that I'm their master.[30]

However, a number of journalists and commentators on the periphery of Hawking's world have made some quite ridiculous extrapolations on this theme. To some, Hawking is a metaphor for his own work, a black hole astronaut himself. When Overbye put this to him, he was understandably ruffled by the suggestion.

'I've always found I could communicate,'[31] he snapped back, and went for Overbye's toes.

Black hole astronaut or not, the amount Hawking travelled during the seventies was increasing each year. In the winter of 1976 he undertook an American tour, taking in talks at important conferences in Chicago and Boston. Even to other scientists who knew him from symposia and conferences around the globe, his speech was all but unintelligible, and when members of the general public and journalists were in attendance they found it almost as difficult to grapple with Hawking's speaking voice as with his subject-matter.

Despite the fact that conference organizers were invariably forewarned of Hawking's disabilities, more often than not there would be no easy access to the stage in the lecture theatre. He would have to make it there without ramps or lifts. On such occasions Hawking's friends and colleagues would come to his rescue, up to six of them manhandling his heavy wheelchair. Although Hawking himself weighed little more than ninety pounds, the chair ran on car batteries which added to the weight and, according to those who have taken part in these exercises, there was always the fear that they would drop him or that he would hurt his neck. One friend has described how he could see Hawking's head bobbing around as six of the biggest scientists in his group lifted the wheelchair five feet up on to the stage, and how he was terrified that one day something would go disastrously wrong, simply because the organizers hadn't thought things through.

Hawking made a great impression during his 1976 trip to the States. The stick-like figure hunched in his wheelchair was, to the vast majority of the audience, mumbling incomprehensibly, appearing to make his pronouncements to a point on the stage six feet in front of him. But, despite this, he was always taken very seriously by those who came to hear him speak. Close colleagues who could understand what he was saying translated for their neighbours as best they could, with one ear concentrating on the mathematics Hawking was describing. Slides and the relief of numerous corny jokes helped, but it was always hard work.

By this time he had completely reversed his ideas about black holes and thermodynamics, the very ideas that had created such arguments a few years earlier. At a talk in Boston, entitled 'Black Holes Are White Hot', he caused a stir with a conclusion refuting Einstein's famous statement 'God doesn't play dice.' 'God not only plays dice,' Hawking proclaimed, 'he sometimes throws them where they can't be seen.'

Interviewers were queueing up to speak to Hawking. In January 1977, the BBC broadcast a programme called *The Key to the Universe*, with an accompanying book, by Nigel Calder. The programme was in large part devoted to Stephen Hawking's latest work, and profiled the man and his efforts to unify relativity and quantum mechanics – 'the key to the Universe' of the title. For the first time, the general public were exposed to the thirty-five-year-old Dr Stephen Hawking, FRS, and the facts of his disability as well as his work. It had the British public watching in their millions.

From 1977, publicity surrounding Hawking and his achievements began to escalate on a local, national and global scale. Between reports of punks signing record contracts in front of Buckingham Palace and growing excitement over the Queen's Jubilee that coming summer, there were mutterings in the Cambridge press about the odd fact that this famous scientist, a member of The Royal Society and black hole celebrity, appearing on television and with his face in the papers on an increasingly regular basis, did not hold a professorial position at Cambridge University.

There were muted suggestions that perhaps the University was disinclined to give the severely disabled scientist a professorship because he might not live too long. By March 1977, however, the University had decided to offer him a specially created Chair of Gravitational Physics, which would be his for as long as he remained in Cambridge; the same year he was awarded the status of professorial fellow at Caius, a separate professorship bestowed by the College authorities.

The awards and honours continued to flood in. Robert Berman, Hawking's undergraduate supervisor at Oxford, had recommended him as an Honorary Fellow of University College. In his letter to the General Purposes Committee, he said:

The current issue of *Who's Who* shows some of his achievements, but cannot keep pace with the rate of award of honours.

I can't imagine that the College has ever produced a more distinguished scientist, and it would bring us honour if our association with his career were made manifest (the outside world assumes he is entirely a Cambridge product).

It might seem surprising to ask to consider someone not yet 35 as an Honorary Fellow, but there are two reasons for this. First, his distinction is quite exceptional and we don't have to wait for it to be generally recognized that he has made his mark. Hawking is mentioned in practically every article or lecture on black holes. His book (*The Large Scale Structure of Spacetime*) was what every cosmologist was waiting for.

Secondly, Hawking is gravely ill and is confined to a wheelchair with a type of creeping paralysis which normally cuts the lives of its victims very short. He is in an appalling physical state but his mind functions normally.

I hope that it won't be felt that we must wait to see whether he actually gets a Nobel Prize!

Berman thought that he may have to argue his case further. He was subsequently staggered when the recommendation was accepted without a single objection at the Committee's first meeting.

The graffiti-daubing sluggard who, at Oxford University only sixteen years earlier, had spent more time drinking than working had come a very long way.

11

Back to the Beginning

By the end of 1974, Hawking's work on black holes had shown that, using the general theory of relativity alone, the equations said that the surface area of a black hole could not shrink – but adding in the quantum rules to the equations revealed that they could not only shrink, but would eventually disappear in a puff of gamma-radiation. His earlier work with Penrose had shown that, using the general theory of relativity alone, the equations said that the Universe must have been born out of a singularity, a point of infinite density and zero volume, at a time some 15 billion years ago. It was natural that the next scientific question Hawking asked himself was what would happen to this prediction if the quantum rules were added to that set of equations.

This was no easy question to answer. Physicists had been trying to combine quantum theory and relativity theory into one complete, unified theory ever since the quantum revolution in the 1920s; Einstein himself spent the last twenty years of his working life on the problem, and failed to come up with a solution. Indeed, a full theory of quantum gravity still eludes the mathematicians. But by restricting himself to the specific puzzle of how relativity and quantum mechanics interacted at the beginning of time, Hawking was able to make progress, to such an extent that by the early 1980s he was posing the question of whether there ever had been a beginning to time at all. To understand how he arrived at this startling hypothesis, we have to look again at the quantum theory, in a variation developed by the great American physicist Richard Feynman. It is known as the 'sum-over-histories' or 'path integral' approach.

The essential features of quantum mechanics are demonstrated most clearly in what is known as 'the experiment with two holes'. In such an experiment, a beam of light, or a stream of electrons, is directed through two small holes in a wall and on to a screen on the other side. The version using light is known as Young's experiment, and may be familiar from school physics. What happens is that the pattern of light on the screen forms a characteristic arrangement of dark and light stripes, caused when the electromagnetic waves passing through each of the holes interfere with each other. Where the two sets of waves add together, there is a bright stripe; where they cancel each other out, the screen is dark.

This interference is easy to understand in terms of waves. You can get exactly the same effect by making waves in a tank of water and letting them pass through two slits in a barrier. But it is much harder to understand how electrons, which we are used to thinking of as hard particles like tiny snooker balls, can behave in the same way. Yet they do.

What is even more strange is that the *same* pattern of dark and light stripes slowly builds up on the screen (which can be almost exactly the same as a TV screen) when electrons are fired through the holes one at a time. Why should this be strange? Think about what happens when electrons are fired through just one hole. Instead of a striped pattern on the screen, there is just a bright patch behind the hole. This is indeed what we see if we block off either of the two holes and fire the electrons through. 'Obviously', each electron can go through only one hole. But when both holes are open, even with electrons fired one at a time through the experiment, we do *not* see just two patches of brightness behind the holes, but the characteristic stripy pattern of Young's experiment.

This is the clearest example of the wave–particle duality (see Chapter 2) which lies at the heart of the quantum world. When each electron arrives at the screen, it makes a pinpoint of light, just as you would expect from the arrival of a tiny 'snooker ball' particle. But when thousands of those points of light are added together, they produce the striped pattern corresponding to a wave passing through both holes at once. It is as if each individual electron is a wave that passes through both holes simultaneously, interferes with itself, decides which bit of the striped pattern it belongs in, and heads off there to arrive as a particle that makes a pinpoint of light.

Don't worry if you find this incomprehensible. Niels Bohr, one of the physicists who pioneered the quantum revolution, used to say that 'anyone who is not shocked by quantum theory has not understood it', while Feynman, probably the greatest theoretical physicist since the Second World War, went even further, and was fond of saying that *nobody* understands quantum mechanics. The important thing is not to understand how such a strange behaviour as wave–particle duality can occur, but to find a set of equations that describe what is going on and make it possible for physicists to predict how electrons, light waves, and the rest will behave. The sum-over-histories approach was Feynman's contribution to this more pragmatic form of 'understanding' at the quantum level, and in the late 1970s Hawking applied it to the study of the Big Bang.

Feynman said that, instead of thinking of an object such as an electron as a simple particle that follows a single route from A to B (for example, through one of the two holes in Young's experiment), we have to regard it as following every possible path from A to B through spacetime. It would be easier for a 'classical' particle to follow some paths (some 'histories') than others, and this is allowed for in Feynman's equations by assigning each path a probability, which can be calculated from the quantum rules.

These probabilities can interfere with the probabilities from neighbouring 'world lines', as they are called, rather like the way ripples on the surface of a pond interfere with one another. The actual path followed by the particle is then calculated by adding together all the probabilities for individual paths (which is why this is also known as the path integral approach).

In the vast majority of cases, the various probabilities cancel each other out almost entirely, leaving just a few paths, or trajectories, that are reinforced. This is what happens for the trajectories corresponding to an electron moving near the nucleus of an atom. The electron is not allowed to go just anywhere, because of the way the probabilities cancel. It is only allowed to move in one of the few orbits around the nucleus where the probabilities reinforce one another.

The experiment with two holes is unusual because it offers the electrons a choice of two equally probable sets of trajectories, one through each hole, and this is why the basic strangeness of the quantum world shows up so clearly in this example. Only Hawking, though, had the chutzpah to apply the path integral approach to calculating the history, not of an individual electron, but of the entire Universe; but even he started out in a smaller way, with black hole singularities.

When a black hole evaporates, what happens to the singularity inside it? One simple guess might be that in the final stages of the evaporation the horizon around the hole vanishes, leaving behind the naked singularity that nature is supposed to abhor. In fact, though, the equations developed by Hawking in the early 1970s to describe exploding black holes could not be pushed to such extremes. Strictly speaking, they could only be applied if the mass of the black hole were still a reasonable fraction of a gram – almost big enough to be weighed on your kitchen scales. The best guess that Hawking, or anyone else, could make in 1974 was that when a black hole has evaporated to this point it would completely disappear, taking the singularity with it. But this was only a guess, based on some general quantum principles.

These principles are aspects of the basic uncertainty principle. Just as there is a fundamental uncertainty about the energy content of the vacuum, so there is a fundamental uncertainty about basic measures such as length and time. The size of these uncertainties is determined by Planck's constant, which gives us basic 'quanta' known as the Planck length and the Planck time.

Both are very small. The Planck length, for example, is 10^{-35} of a metre, far smaller than the nucleus of an atom. According to the quantum rules, not only is it impossible in principle ever to measure any length more accurately than this (we should be so lucky!), but there is no meaning to the concept of a length shorter than the Planck length. So if an evaporating black hole were to shrink to the point where it was just one Planck length in diameter, it could not shrink any more. If it lost more energy, it could only disappear entirely. The quantum of time is, similarly, the smallest interval of time that has any meaning. This Planck time is a mere 10^{-43} of a second, and there is no such thing as a shorter interval of time. (Don't worry about the exact size of these numbers; what matters is that although they are exceedingly small, they are *not* zero.) Quantum theory tells us that we can neither shrink away a black hole to a mathematical point, nor look back in time literally to the moment when time 'began'. Even if we pushed the Big Bang model to its most extreme limit, we would have to envisage the Universe being created with an 'age' equal to the Planck time.

In both cases, quantum mechanics seems to remove the troublesome singularities. If there is no meaning to the concept of a volume with a diameter less than the Planck length, then there is no meaning to the concept of a point of zero volume and infinite density. Quantum theory is telling us that although the densities reached inside black holes, and at the birth of the Universe, may be staggeringly high by any human measure, they are not infinite. And if the infinities and singularities can be removed, then there is at least a hope of finding a set of equations to describe the origin (and, it turns out, the fate) of the Universe. Having started out in 1975 from the puzzle of what happens in the last stages of the evaporation of a black hole, by 1981 Hawking was ready to unveil his new ideas, incorporating Feynman's sum-over-histories version of quantum mechanics, to explain how the Universe had come into being. The place he chose for the unveiling was – the Vatican.

In fact, the choice of venue was not entirely Hawking's whim. It happened that the Catholic Church had invited several eminent cosmologists to attend a conference in Rome in 1981, to discuss the evolution of the Universe from the Big Bang onwards. By the 1980s, the Church was much more receptive to scientific teaching than it had been in the days of Galileo, and the official view was that it was quite OK for science to investigate events since the Big Bang, leaving the mystery of the moment of creation in the hands of God.

Fortunately, perhaps, Hawking's investigation of the moment of creation was still couched in rather abstruse mathematical language when he presented it to that conference. Since then, however, he has developed the ideas in a more accessible way (most notably with the help of James Hartle, of the University of California). It doesn't take much intuition to guess that the Pope would probably not approve of the fully developed version of Hawking's ideas, which seems to do away entirely with a role for God.

What Hawking has tried to do is to develop a sum over histories describing the entire evolution of the Universe. Now this is, of course, impossible. Just one history of this kind would involve working out the trajectory of every single particle through spacetime from the beginning of the Universe to the end, and there would be a huge number of such histories involved in the 'integration'. But Hawking found that there is a way to simplify the calculations, provided the Universe has a particularly simple form.

Quantum theory comes into the calculations in the form of the sum over histories. General relativity enters in the form of curved spacetime. In Hawking's models, a complete curved spacetime that describes the entire history of a model universe is equivalent to a trajectory of a single particle in Feynman's sum over histories. General relativity allows for the possibility of many different kinds of curvature, and some sorts of curvature turn out to be more probable than others.

If the Universe is like the interior of a black hole, with spacetime closed around it, then we can imagine, in the standard picture of the Big Bang, that everything (including space) expands outwards from the initial singularity, reaches a certain size, and then collapses back into a mirror image of the Big Bang, the so-called 'Big Crunch'. In this picture, there is a beginning of time in the initial singularity, and an end of time in the final singularity. Hawking calls the beginning and end of time 'edges' to this model of the Universe – such a model has no edge in space, because space is folded round into a smooth surface like the surface of a balloon, or the surface of the Earth; but there is an edge in time in the beginning, when the Universe appears as a point of zero size.

Hawking wanted to remove the edge in time, as well as the edge in space, to produce a model of the Universe which has no boundaries at all. He found that, without having to go into the detail of calculating every trajectory of every particle through spacetime, the general rules of the sum-over-histories approach as applied to families of curved spacetimes said that a certain kind of curvature is much more likely than any other, if the no-boundary condition applies.

Hawking stresses that this no-boundary condition is, as yet, just a guess about the nature of the Universe, but it is a guess that leads to a powerful image of reality. This is the cosmological equivalent of saying that the path integral approach tells us that an electron can follow only certain orbits around a nucleus; the Universe has only a limited number of life cycles to choose from, and they all look much the same.

The best way to picture these models is by an extension of the idea of the Universe being represented by the surface of a balloon. In the old picture, this surface represents space, and the evolution of the Universe from bang to crunch is represented by imagining the balloon being first inflated and then deflated. In the new picture, however, the spherical surface represents both space and time, and it stays the same size – much more like the surface of the Earth than the surface of an expanding balloon. So where does the observed expansion of the Universe come into this model?

Now, says Hawking, we have to imagine the Big Bang as corresponding to a point on the surface of the sphere, at the North Pole. A tiny circle drawn around that point (a line of latitude) corresponds to the size of the space occupied by the Universe. As time passes, we have to imagine lines of latitude being drawn further and further away from the North Pole, getting bigger (showing that the Universe expands) all the way to the equator. From the equator down to the South Pole, the lines of latitude get smaller once again, corresponding to the Universe shrinking back to nothing at all as time passes.

We still have an image of the Universe being born in a superdense state, evolving, and shrinking back into a superdense state, but there is no longer a discontinuity in time, just as there is no edge of the world at the North Pole. At the North Pole, there is no direction north, and every direction points south. But this is simply due to the geometry of the curved surface of the Earth. In the same way, at the Big Bang there was no past, and all times lay in the future. And this is simply due to the geometry of curved spacetime. The whole package of space and time, matter and energy, is completely self-contained.

A rather nice way to understand what is going on is to imagine you are standing a little way from the North Pole, and start to walk due north. Even though you keep walking in a straight line, you will soon find that you are walking due south. In the same way, if you had a working time-machine, and started travelling backwards in time from some moment just after the Big Bang, you would soon find that you were travelling forwards in time, even though you had not altered the controls of the time-machine. You just cannot get back to a time before the Big Bang (strictly speaking, before the Planck time), because there simply is no 'before'.

In *A Brief History of Time*, Hawking spelt out the implications for religion. He leaves his colleagues in no doubt that he is, at the very least, an agnostic, and finds strong support for this belief in his cosmological studies:

So long as the universe had a beginning, we could suppose it had a creator. But if the universe is really completely self-contained, having no boundary or edge, it would have neither beginning nor end: it would simply be. What place, then, for a creator?[1]

But even without a creator there were still problems to be solved. Already, in 1981, the attention of Hawking and other theorists was focusing on the next question – how did a tiny seed of a Universe get blown up to the enormous size that we see today?

The puzzle of how the Universe has got to be as big as it is today had itself loomed larger and larger during the 1970s. When everybody thought that the Big Bang theory was just a model to play with, they didn't worry too much about the details of how it might work. But as evidence built up that this model provides a very good description of the real Universe, it became increasingly important to explain *exactly* what makes the model, and the Universe, tick.

There were two problems that cosmologists were simply unable to answer in the 1970s. First, why is the Universe so uniform – why does it look the same (on average) in all directions of space, and why, in particular, is the temperature of the microwave background exactly the same in all directions? Secondly, the Universe seems to be delicately balanced on the dividing line between being closed, like a black hole, and open, so that it will expand forever. In terms of the curvature of space, the Universe is remarkably flat. Why is this?

On the basis of general relativity alone, there seems to be no reason why it could not have been, for example, much more tightly curved, in which case the Universe would have expanded only a little way out of the Big Bang before recollapsing, and there would have been insufficient time for stars, planets and people to evolve. Cosmologists suspected that the smoothness and flatness of the Universe were telling us something fundamental about the nature of the Big Bang, but nobody could see just what that might be until a young researcher at Cornell University, Alan Guth, came up with a new idea.

Guth's proposal goes by the name 'inflation', and stems from quantum physics. He suggested that in the first split-second after the beginning, the vacuum of the Universe existed in a highly energetic state, as allowed by the quantum rules, but unstable. The high-energy state is analogous to a container of water cooled, very slowly and carefully, to *below* 0 °C. Such supercooling is possible, if the water is cooled very carefully, but the result is unstable. At a slight disturbance, the water will freeze into ice, and as it does so it gives up energy (exactly the same amount of energy that is needed to melt an ice-cube, at 0 °C, is released when the same amount of water freezes).

This is where the ice analogy breaks down slightly, for when the Universe cooled from the excited vacuum state to the stable vacuum that we know today, so much energy was released that it became superhot, not icy, and for a time it expanded superfast. In a tiny fraction of a second, a region of space far smaller than a proton (but packed full of energy) must have been inflated, according to this theory, into a volume about the size of a grapefruit. At that point the inflation was exhausted, and the grapefruit-sized fireball began the steady expansion associated with the standard model of the Big Bang, growing over the next 15 billion years to become the entire visible Universe.

According to inflation theory, the Universe is so uniform because it has grown out of a seed so small that there was literally no room inside it for irregularities. And the equations also tell us that the inflation process flattened space. The best analogy for how this works is with the wrinkly surface of a prune, which is very far from flat. When you soak the prune in water, it swells up, expanding so that the surface stretches and the wrinkles are smoothed out. Imagine starting out with a prune smaller than a proton, and expanding it to the size of a grapefruit, and you can see why space is so very flat today.

The inflationary model has been extensively developed since Guth made the original proposal in 1980. Hawking has been involved in filling in details of this work throughout the 1980s, but the main developments have come from a Soviet researcher, Andrei Linde. Some of Linde's early contributions were duplicated independently by Paul Steinhardt and Andreas Albrecht, from the University of Pennsylvania. As we shall see in Chapter 15, the early versions of inflation were overtaken in the 1980s by new insights which provide a spectacular new image of the origin and evolution of not just *the* Universe, but a multiplicity of universes. Hawking played a part in this work, too. From now on, honours and awards would be heaped upon the man to whom the modest recognition offered by the Gravity Research Foundation had been 'very welcome' just a short time before.

12

Science Superstardom

In 1978 Hawking was awarded one the most prestigious prizes in physics, the Albert Einstein Award given by the Lewis and Rose Strauss Memorial Fund, which announced the winner at a gala event in Washington. The citation claimed that Hawking's work could lead to a unified field theory, 'much sought after by scientists',[1] as one Cambridge newspaper put it. The Albert Einstein Award is considered to be the prestigious equivalent of a Nobel prize, and was undoubtedly the most important award Hawking had received up until that time. Journalists began to talk about the possibility of the thirty-six-year-old physicist being next in line for the greatest academic honour of all – an invitation to the Royal Academy of Sciences in Stockholm.

However, there are two reasons why Hawking is unlikely ever to receive a Nobel Prize. First, a cursory glance at the list of winners since the first prizes in 1901 shows very few astronomers. The reason for this, according to one story, is that the chemist Alfred Nobel, who created the awards, decreed that astronomers should be ineligible. Rumour has it that their exclusion was because his wife had an affair with an astronomer, and he subsequently felt only hatred for the whole profession. Despite this, Martin Ryle and Antony Hewish shared the 1974 Nobel Prize for Physics for their work in radioastrophysics and Subrahmanyan Chandrasekhar won it in 1983 for his theoretical studies on the origin and evolution of stars. These were awarded a good seventy years after the founder's death, so perhaps the Academy now views astronomers with greater sympathy.

There is, however, a more important reason for Hawking's absence from the list of winners. One of the Academy's rules states that a candidate may be considered for a prize only if a discovery can be supported by verifiable experimental or observational evidence. Hawking's work is, of course, unproved. Although the mathematics of his theories is considered beautiful and elegant, science is still unable even to prove the existence of black holes, let alone verify Hawking Radiation or any of his other theoretical proposals.

A year after receiving the Albert Einstein Award, Hawking's second book was published by Cambridge University Press: a collection of sixteen articles to commemorate the centenary of Albert Einstein's birth on 14 March 1879. Hawking co-edited the book, entitled *General Relativity: An Einstein Centenary Survey*, with his colleague Werner Israel. When Simon Mitton presented it to a sales conference in January 1979, the sales team, whose job it was to take books out on the road and convince retailers of their merit, was unusually enthusiastic. One of the sales staff said to Mitton, 'That man Hawking – he's amazing, you know. We'll have no trouble selling this. All the quality bookshops will take it, no problem.' He was right. It was snapped up and sold exceptionally well in hardback and even better when later issued as a paperback. Hawking's fame was spreading.

This was also the year that Stephen Hawking finally got his own office at the DAMTP – it came with his appointment as Lucasian Professor. Hawking is well aware of his place in the history of science. He is fascinated by the fact that he was born on the three-hundredth anniversary of Galileo's death on 8 January 1642. That year Isaac Newton was born in Woolsthorpe, a little village in Lincolnshire, and it was Isaac Newton who was appointed Lucasian Professor at Cambridge in 1669, three hundred and ten years before Hawking.

Galileo was considered the greatest of all scientists by Albert Einstein, and Hawking has claimed that he was, in his approach, the first twentieth-century scientist:

He was the first scientist to actually start using his eyes, both figuratively and physically. And, in a sense, he was responsible for the age of science we now enjoy.[2]

Galileo's work led directly to Newton's work and the establishment of classical physics. The work of Einstein, who was born one hundred years before Hawking received the Lucasian chair, turned 'large scale' physics on its head. Subsequently, Hawking has been seen by many as the physicist most likely to succeed in the enormous task of unifying the two supporting pillars of physics, quantum mechanics and relativity. Small wonder Hawking has a strong sense of science history.

At his inauguration as Lucasian Professor, Hawking delivered a memorable lecture, entitled 'Is the End in Sight for Theoretical Physics?', in which he suggested that a Grand Unified Theory describing the fundamental laws of the Universe could be achieved by the end of the century.

It was a stirring and inspiring idea. The audience knew as they streamed out of the hall that if anyone could make that dream come true, it would be the waif-like figure who had earlier sat on the stage before them, crumpled in his motorized wheelchair, delivering powerful statements with his typical confidence.

The appointment as Lucasian Professor of Mathematics at Cambridge University was one of the highlights of Hawking's career. To be professor at one of the oldest and most respected universities in the world is a huge achievement in itself, but to have accomplished such a feat by the age of thirty-seven is remarkable. Newton was Hawking's junior by ten years when he gained the Chair, but in the seventeenth century there were far fewer academics and very little competition for such positions. Newton did also happen to be the youngest ever to be appointed Lucasian Professor at Cambridge.

Easter 1979 saw the birth of Stephen and Jane's third child, a boy they christened Timothy. It was a happy time for the Hawking family. Against all odds, they had overcome tremendous hurdles to achieve great success. Jane had completed her PhD, and was finding a degree of intellectual satisfaction in her teaching job; Professor Hawking was receiving the esteem of his colleagues and growing popular acclaim as 'the new Einstein'. Now there was another Hawking at West Road.

In the larger world outside the cloistered environs of Cambridge academia, the ever-shifting kaleidoscope of life was shaken yet again. Shortly before Timothy Hawking's birth, scientists at the Jet Propulsion Laboratory in Pasadena were surprised to discover, via the deep space probe Voyager 1, that Jupiter had rings like its celestial neighbour, Saturn. Before the year was out, Margaret Thatcher had begun her eleven-year run as Britain's first woman Prime Minister, the Queen's cousin, Lord Mountbatten of Burma, was murdered by the IRA, and American embassy staff and marines were taken hostage in Tehran. Also that year, the Queen's art adviser, Cambridge man Anthony Blunt, was exposed as the 'fourth man'. Russia invaded Afghanistan, Mother Teresa of Calcutta was awarded the Nobel Peace Prize, and John Cleese continued to delight TV audiences by 'not mentioning the war'. One of the year's biggest films was *Apocalypse Now*.

At the turn of the decade, Hawking could look back satisfied with his achievements over the past ten years. The symptoms of ALS had levelled off. His speech was practically unintelligible to all but his close colleagues and family, and he was confined to his motorized wheelchair, but he continued to work and to travel as intensively as he had ever done. His freedom from mundane chores and responsibilities was paying dividends scientifically.

*

From 1980, the system of taking in graduate students to help around the house was replaced by community and private nursing. Jane had help looking after Stephen for a couple of hours in the morning and evening. They could just afford to flesh out the meagre assistance provided by the National Health Service by dipping into monies Hawking had received from the growing number of awards and prizes coming his way and the increased salary from his new appointment.

Stephen and Jane began to cultivate a reputation as socialites and popular hosts on the Cambridge academic scene. Don Page has described Jane as 'a great professional asset to her husband as a hostess'.[3] Dr Berman, Hawking's tutor at Oxford, has said of her, '[Jane is] a remarkable woman. She sees that he does everything that a healthy person would do. They go everywhere and do everything.'[4] The Hawkings were soon at the centre of the social in-crowd at Cambridge. Being Lucasian Professor gave Stephen a huge measure of prestige, both in academic circles and in the broader view of the international intelligentsia. Dinner-parties and social gatherings in West Road and at the DAMTP were frequent events, and guests often included visiting academics as well as members of the University hierarchy. Their interest in classical music was well catered for in Cambridge, and the couple were often to be seen at concerts in the city. They enjoyed going to the theatre and the cinema and dining out, both at home in Cambridge and on visits abroad.

Stephen's obvious handicaps would sometimes cause embarrassment to those who did not know him in restaurants and at various functions to which the couple were frequently invited. Casual onlookers, unaware of the fact that they were in the presence of one of the world's greatest scientists, could be forgiven for thinking that the withered figure slumped in his wheelchair – trying to speak but succeeding only in producing an incomprehensible noise, having to be fed, his head, insufficiently supported by atrophied neck muscles, rolling forward, chin on chest – was a hopelessly crippled and pathetically disabled man, perhaps mentally as well as physically handicapped. Nothing could be further from the truth. On the subject of his disability, Hawking told an interviewer at the time:

> I think I'm happier now than I was before I started. Before the illness set in I was very bored with life. I drank a fair amount, I guess, didn't do any work. It was really a rather pointless existence. When one's expectations are reduced to zero, one really appreciates everything that one does have.[5]

On another occasion he said, 'If you are disabled physically, you cannot afford to be disabled psychologically.'[6]

Jane echoed this view, with a typically forthright and optimistic approach to life. 'We try to make the most of every moment,'[7] she told one interviewer.

A *Sunday Times* journalist once asked him whether he ever got depressed because of his disability.

'Not normally,' he replied. 'I have managed to do what I wanted to do despite it, and that gives me a feeling of achievement.'[8]

Another asked what was his biggest regret about contracting his illness.

'Not being able to play physically with my children,' he said.[9]

Some years earlier, Hawking had entered into a protracted fight with the University authorities over improved access for him in the DAMTP building. The row was about who was to pay for a ramp to be installed. Hawking eventually won and also managed to persuade the authorities to lower the kerbs in the vicinity of Silver Street to ease his journey from West Road. Such clashes put Hawking in a fighting mood about the needs of the disabled, and he has been crusading for various causes ever since.

He took on Cambridge City Council over access to public buildings and won. After a long-drawn-out argument and an exchange of increasingly abrasive letters, kerbs were lowered in a number of vital places and ramps installed in various buildings. One particular dispute concerned a public building called Cockcroft Hall, used as a polling station during local elections. After polling day, Hawking complained to the council that it was practically impossible for the severely disabled to enter the building in order to vote. The council authorities tried to argue that Cockcroft Hall was not actually a public building and did not therefore come under the Disabled Persons Act of 1970. Because of Professor Hawking's involvement, the local press became interested in the issue and subsequently ran a series of articles highlighting the problems faced by the disabled in Cambridge. The City Council backed down.

Towards the end of 1979, the Royal Association for Disability and Rehabilitation nominated Hawking for 'Man of the Year' and his efforts in fighting for the rights of handicapped people were again noted by the local press, which held him up as a champion of their cause. Hawking himself has ambivalent feelings on this issue. On the one hand, he wants to do what he can for other handicapped people, for, being disabled himself, he knows and fully understands the problems faced by the handicapped. He has a stubborn streak and definite strains of a rebellious nature, partly cultivated by his circumstances, which give him an appetite for dispute. He loves nothing more than a good argument, whether it is about cosmology, socialism or the rights of the disabled. On the other hand, Hawking has always made a conscious effort to detach himself from his condition. He has absolutely no interest in learning more about his illness or over-emphasizing his disability.

One interviewer asked him if he regretted not using his intellectual powers to help find a cure for his illness. He replied that he would have found that too upsetting. He is a physicist, not a medical man, and knowing the gruesome details would, he feels, be totally unproductive. Hawking is, of course, very happy that others are working on a cure for ALS, but he does not wish to know how the research is going. He just wants to be told when they have made a breakthrough.

All this led to what was perceived at the time to be a strangely ambivalent attitude to the problems faced by the disabled. Critics began to complain that he was not doing enough, that his growing celebrity was a perfect platform for him to be heard above the crowd. As time has gone on, Hawking has indeed become more active, but the simple fact is that he hardly needs do anything because, just by staying alive and continuing to work at the intense rate he and the world have grown used to, he is an inspiration to handicapped people everywhere.

In a recent speech at an occupational science conference at the University of Southern California, he certainly made every effort to raise his voice above the crowd:

> It is very important that disabled children should be helped to blend with others of the same age. It determines their self-image. How can one feel a member of the human race if one is set apart from an early age? It is a form of apartheid. Aids like wheelchairs and computers can play an important role in overcoming physical deficiencies; the right attitude is even more important. It is no use complaining about the public's attitude about the disabled. It is up to disabled people to change people's awareness in the same way that blacks and women have changed public perceptions.[10]

Having got a taste for it, Hawking did not restrict his campaigning to the problems of the disabled. He was beginning to show a growing interest in saying his piece about a number of wide-ranging socio-political issues. He led a campaign to change the ruling prohibiting the admission of women students into Caius College, a row which lasted the best part of a decade. He and Jane continued to be paid-up members of the Labour Party, and he was becoming increasingly vocal on social issues such as the plight of the poor and the state of the environment. He has joked that he is a 'right-wing socialist', but his attitudes towards concerns ranging from the Falklands War to nuclear disarmament show definite leanings towards a brand of liberalism prevalent in the Hawking household of his early years.

When accepting an award sponsored by a US defence contractor, he lectured the executives of the company gathered at the ceremony on the senselessness of nuclear weapons:

> We have the equivalent of four tons of high explosives for every person on earth. It takes half a pound of explosive to kill one person, so we have 16,000 times as much as we need. We must understand that we are not in conflict with the Soviets, that both sides have a strong interest in the stability of the other side. We ought to recognize that fact and cooperate, rather than arm ourselves against each other.[11]

Apart from getting his own office, life at the DAMTP had changed little upon his appointment as Lucasian Professor. Silver Street is a narrow winding lane off King's Parade in the centre of Cambridge. The sign for the Department of Mathematics and Theoretical Physics is unobtrusive to the point of near-uselessness – visitors frequently find themselves unable to find the entrance unassisted. When finally discovered, the sign indicates an archway leading on to a cobbled courtyard. A number of cars are parked around the perimeter and there are stacks of bicycles, three deep, propped up against the stonework. At the far end of the courtyard is a red door with a glass window and, on a wall to one side, is a brass plate announcing the Department in a clearer and more elegant fashion.

Inside, a linoleum-floored hallway leads to a large, scruffy common-room. Tables and low, soft chairs are randomly distributed around the room, left where positioned by their most recent occupants. The walls are painted grey, and the whole atmosphere is one of academic drabness, slightly neglected, workaday. From the common-room, doors lead off to a number of offices. The one Hawking shared with a former student, Gary Gibbons, sports a sticker which says 'Black Holes are Out of Sight'. The door to his new office has a typically self-mocking addition pasted at head height: 'QUIET PLEASE, THE BOSS IS ASLEEP'.

Hawking's office has changed little since he took it over in 1979. It is relatively small and dominated by a desk set two-thirds of the way back from the door. The walls are lined with book-shelves, and to one side of the desk sits a set of gadgets. The first is a telephone especially adapted with a microphone and loudspeaker so that he can use it without having to hold the handset. Next to that is another device – a page-turner which automatically leafs through any book placed on a raised platform, operated at the touch of a button. Once an assistant has positioned a book for him and set the fasteners, Hawking can find any place in the text he wishes to read. Complications arise if he wants to consult a paper or read a magazine because the machine cannot handle them. On these occasions, the article has to be xeroxed and laid out on the desk for him. On the desk, next to framed pictures of the family, is a computer, augmented by the addition of two levers which operate a cursor on the screen. This replaces the normal keyboard, and doubles as a 'blackboard' and word-processor.

There is a relaxed atmosphere in the Department. Perpetuating the tradition of several decades, everyone meets twice daily for morning coffee and afternoon tea. At these gatherings the talk centres on the day's work. Spending five minutes in the DAMTP's common-room reveals an obvious fact: physicists love to talk shop. The students treat Hawking with playful irreverence; there is no standing on ceremony or élitism here. When the writer Dennis Overbye visited Hawking at the DAMTP he came across a group of students huddled around a Formica-topped table in the common-room. 'In age, dress, pallor and evidence of nutritional deficiency, they resembled the road-crew of a rock-and-roll band'[12] is how he described them. Hawking mucks in with them, cracking corny, undergraduate jokes. Following an old tradition, if they hit on a bright idea during the course of their discussions they write out mathematical descriptions on the table-tops. 'When we want to save something we just xerox the table,'[13] Hawking told Overbye.

Hawking's administrative duties extended to running the small relativity group, which consisted of a dozen or so research assistants of wide-ranging nationality, and the supervision of a handful of PhD students. Apart from these responsibilities, the professorship allowed him to carry on with what he had previously devoted so much time to – thinking.

At home, Hawking's schedule was a hectic one. Hardly a week would go by without a visit from a foreign colleague. It was now his responsibility to organize symposia and lectures given by physicists interested in visiting Cambridge. Hawking's relativity group at the DAMTP was seen as being at the cutting edge of research, and there was no shortage of scientists interested in sharing their latest work with the Cambridge team.

Hawking had, by this time, established an exhausting work routine at the DAMTP, one that has changed little to this day. He rose early, but it could take up to two hours for him to get ready to leave the house, arriving at his office by 10 a.m. The journey from West Road took no more than ten minutes and was usually spent in conversation with one of his

PhD students or research assistants. After checking the post with his secretary, he usually spent the morning working at his computer or reading articles or papers written by others in the field. At 11 a.m. sharp he would wheel himself off to the common-room where an assistant helped him with drinking his coffee, lifting the cup to Hawking's mouth. He then often spent some time conversing, as best he could, with the students and research assistants, before returning to his office until lunchtime to make and receive telephone calls and answer correspondence.

At 1 p.m. precisely he would set off for luncheon at Caius College. Usually accompanied by an assistant, he would set the control toggle of the wheelchair to full throttle and head off towards King's Parade, passing by King's College Chapel and the Senate House, his assistant having to break into a trot to keep up. Hawking loves this city in which he has spent most of his life. The grandeur of its architecture and the atmosphere of intense intellectual activity pervading the place are very important to him. Accompanied on this journey by one writer, he gave the interviewer a history lesson, tinged with his characteristic brand of irony:

When Dr Caius reopened Gonville College in the sixteenth century, he built three gates. You entered through the Gate of Humility, you passed through the Gate of Wisdom and Virtue, and you left through the Gate of Honour. The Gate of Humility has been torn down. It's not needed any more.[14]

After lunch each weekday, Hawking headed off back to the DAMTP to work until tea-time. At 4 p.m. the usually silent common-room would erupt with the noise of those who work there. Tea was drunk as a number of animated conversations took place in small groups. Then, as now, Hawking usually sat in one corner of the room. He rarely says more than a few sentences during the course of tea, but when he does speak, people listen. One student has remarked that more can be gained from a few of Hawking's crisp, precise statements than from a whole lecture by anyone else.

His students usually came to see him in the late afternoon. They would sit beside him at his desk or perch alongside the desk-top computer screen. With the sheets of equations they had been working on spread before them, Hawking would survey their efforts, and make a few clipped suggestions. His close associates, his research assistants, then fleshed out his comments and helped the PhD students unravel problems and expand on the professor's suggestions.

After tea, Hawking usually worked until 7 p.m. He would then wheel his chair out of the building and re-run the morning's journey in reverse. Some evenings he chose to dine with the other dons and professors at high table in College. On such occasions he would be obliged to dress in his professorial gown. At other times he stayed at home with Jane and the children, or the couple would go out to eat at a Cambridge restaurant while one of Hawking's helpers babysat.

As his celebrity grew, the amount of time Hawking spent travelling abroad increased further. During the early 1980s he made several trips to America each year, and attended numerous conferences and lectures in Europe and other parts of the globe. Roger Penrose has recalled that nothing would stop Stephen making trips to far-flung destinations and that he would try to attend every important conference, no matter where it was held. At one conference held in Belgium, he almost missed the plane home from Brussels because the cab-driver taking him and Penrose to the airport got lost. Arriving at the airport, with the plane on the tarmac ready to leave, Penrose had to race along ramps and through airport buildings with Hawking's wheelchair whirring along at full throttle beside him. They just made it in time, boarding the aircraft minutes before take-off.

Jane began to travel abroad less frequently so that she

could look after the growing family in Cambridge. The responsibility of nursing Hawking on foreign visits increasingly fell to his research assistants and close colleagues. Friends like Penrose would help out as best they could and travel with him when they were attending the same conference, but by this time one of his students would always have to go with him everywhere he went. Whenever possible Hawking tried to stretch the budget in order to finance a nurse to accompany him and his academic assistant. In this respect things were easier after he became Lucasian Professor, but even so, academic institutions do not like to splash money around. By this time, however, Hawking had become sufficiently important, and his case exceptional enough, for rules to be bent somewhat.

If they did not travel with him to destinations all over the world, the family were certainly never forgotten. Penrose remembers one incident when their return flight was delayed and they had to spend several hours in an airport lounge. Hawking had spotted a cuddly toy in the display window of one of the shops. He told his friend that he wanted that particular toy to take home for Lucy. Commandeering Penrose to buy it for him, Hawking spent the rest of their wait with a large, pink fluffy animal perched on his lap, practically swamping his wasted body. Lucy was of course delighted with the gift.

When Hawking attended the ground-breaking cosmology conference organized by the Pontifical Academy of Sciences in the Vatican in 1981 (see Chapter 11), Jane went with him. The conference delegates and their partners spent a week in Rome. On a number of evenings Stephen and Jane went out to restaurants, often sharing their table with Dennis Sciama and his wife Lydia, as well as other friends who were also attending the conference. Jane remembers the trip as a happy time for the two of them. Between meetings and discussions, Stephen tried to make time for sightseeing, one of his favourite pastimes.

In his address to the conference, the Pope warned the physicists against delving too deeply into the question of how or why the Universe began, reminding them that this was solely a matter for theologians. He went on:

> Any scientific hypothesis on the origin of the world, such as that of the primeval atom from which the whole of the physical world derived, leaves open the problem concerning the beginning of the Universe. Science cannot by itself resolve such a question; what is needed is that human knowledge that rises above physics and astrophysics which is called metaphysics; it needs above all the knowledge that comes from the revelation of God.[15]

Hawking sat impassively in his wheelchair listening as Pope John Paul II told them that he saw nothing wrong with modern cosmology and even believed that there may be some substance to the idea of the Big Bang. But that, he said, was where the line of demarcation should be drawn, and cosmologists should not try to look beyond it. Some of the older scientists in attendance were reminded of another conference held at the Vatican in 1962, when the then Pope, John XXIII, declared that he hoped they would all follow the example of Galileo! It was at the 1981 Vatican Conference that Hawking announced his controversial 'no-boundary' theorem and the religious connotations accompanying it. It was received enthusiastically by the audience, but what the Pope thought of the idea has not been reported. If nothing else, Hawking certainly has a highly developed sense of occasion.

After the conference, the visiting physicists and their spouses were invited to an audience with the Pope at his summer residence, Castel Gandolfo. The building itself is unimposing but possesses a simple beauty. Visitors pass through the little village surrounding the grounds and up to the house via a long driveway. The scientists from the Vatican were not the only guests of the Pope that afternoon, and security at Castel Gandolfo (and indeed, in the Vatican City) was as tight as could be expected. That year, 1981, will surely be remembered as the year of assassination attempts.

Six months earlier, ex-Beatle John Lennon had arrived at his apartment in the Dakota Building in New York with his wife, Yoko Ono. Moments later he was senselessly gunned down by a psychopath, Mark Chapman, and millions of fans the world over were shaken at what they saw as the end of an era. In March 1981, the recently inaugurated President Reagan had been hit in the chest by a .22 bullet, and less than two months later Pope John Paul II himself had nearly died when he was struck by four bullets from a 9 mm Browning, one of which lodged in his lower intestine. The audience at Castel Gandolfo was the Pope's first public appearance since the incident in St Peter's Square that had almost taken his life.

Following a private meeting with the physicists the Pope gave a speech in the main reception room, after which his guests were introduced to him in person as he sat on a raised chair upon a dais guarded by Papal security. The visitors entered from one side of the platform, knelt before the Pontiff, exchanged a few muttered words, then left on the far side of the stage. When it was Hawking's turn, he wheeled on to the stage and up to the Pope. The other guests watched as the man who, only days earlier, had talked of the 'no-boundary' concept and the fact that there could be no need for a Creator came face to face with the leader of the Catholic Church and, for millions, God's representative on Earth. Everyone, believer and cynic alike, was curious to know what would be said. However, no one in the room could have been more surprised by what happened next. As Hawking's wheelchair came to a halt in front of the Pope, John Paul left his seat and knelt down to bring his face to Hawking's level.

The two men talked for longer than any of the other guests. Finally the Pope stood up, dusted down his cassock and gave Hawking a parting smile, and the wheelchair whirred off to the far side of the stage. There were a number of offended Catholics in the hall that afternoon, misinterpreting the Pope's gesture as undue respect. Many of the non-scientists present were unfamiliar with Hawking's latest proposals, but his reputation as a scientist with irreligious views was well known. They simply could not understand why the Pope should kneel before him; to them, Hawking's opinions were at the opposite end of the spectrum from orthodox Catholic doctrine. Why had John Paul not taken more interest in them, the faithful?

Back at the DAMTP, work continued as usual. Hawking's third book for Cambridge University Press was published soon after his return. However, this time things did not run so smoothly, and there was a whole series of arguments between Hawking and Simon Mitton before the book saw the light of day. It was to be called *Superspace and Supergravity*, aimed at about the same level as *The Large Scale Structure of Spacetime*, and was expected to sell in similar numbers to its predecessor – between five and ten thousand copies over a period of years. The source of the dispute between Hawking and the publishers was the choice of cover for the book.

Hawking wanted a drawing from the blackboard in his office to be photographed and used on the dust-jacket of the hardback edition, as well as on the cover when the book was issued in paperback. The trouble began when Simon Mitton realized that the picture, a bizarre cartoon covered with in-jokes and witticisms done by a group of colleagues after a recent conference at the DAMTP, had been drawn in colour and required full-colour printing. Hawking would not consider a black-and-white photograph of the illustration and was absolutely adamant about using a full-colour representation.

Cambridge University Press insisted that they had never done a four-colour cover for a book such as Hawking's, which, even accepting his international fame as a scientist, would not sell enough copies to warrant the expense. The cover, they stated, would make absolutely no difference to the number of copies the book sold. At this point Hawking saw red and declared that unless they agreed to use his cover he would withdraw the book completely. After a hastily convened editorial meeting, Mitton capitulated, but he was right – *Superspace and Supergravity* sold marginally less than *The Large Scale Structure of Spacetime.*

While the dispute with Cambridge University Press was in full flow and Hawking miraculously found time to work, travel, see his family and engage in bureaucratic wrangles with the city authorities and University, the world at large was going through its usual turmoils. Riots hit British cities; there was intensified fighting in Beirut; and President Anwar Sadat of Egypt was brutally assassinated on 6 October during a military parade in Cairo. In December, doctors in the USA were alerted to a new deadly illness which appeared to attack the body's immune system. But the news in 1981 was not all bad. In July, an estimated 700 million TV viewers tuned in to see Prince Charles marry Lady Diana Spencer in St Paul's Cathedral; England claimed a remarkable cricketing victory against Australia; and the New Year Honours List announced at the end of December included a wheelchair-bound Cambridge physicist who had pioneered important work on black holes – Stephen Hawking was made a Commander of the British Empire by Queen Elizabeth II.

As the 1980s progressed, awards and honours continued to be bestowed on Hawking. In 1982 alone he was made Honorary Doctor of Science by no fewer than four universities: the University of Leicester in Britain, New York, Princeton and Notre Dame universities in the USA.

The interest of the media intensified as Hawking's recognition grew. In 1983, a BBC *Horizon* programm profiled him at work at the DAMTP. For the first time the British public was given a chance to see Professor Hawking whirring around

Cambridge in this wheelchair, talking in his strangely contorted way with his students and co-workers, at home in West Road with Jane and the children, and attending official functions. The public was captivated. One magazine article after another appeared in rapid succession. The London *Times* and *Telegraph* newspapers ran pieces about him, and in-depth interviews turned up in the *New York Times*, *Newsweek* and *Vanity Fair*. A few short years into the decade, and 'black hole' and 'Stephen Hawking' had become synonymous in the eyes of the media and the general public.

Hawking has never been a man to shy away from publicity and he thoroughly enjoyed his growing fame. However, fame alone does not pay the bills, and in the early eighties there were intensifying financial pressures on the Hawking household. A professor's salary is not large compared with equivalent positions in industry or commerce, and occasional monies from prizes and awards were erratic and usually too small to make any real difference. With the strain of running a home and maintaining her own career, Jane was finding that the little nursing help they could afford was growing increasingly inadequate. She desperately needed more private nursing assistance, and that would be expensive.

That was not all. They had managed to finance their eldest son Robert's education at the fee-paying Perse School in Cambridge since the age of seven. He had been highly successful academically, and was scheduled in a few short years to go to university. Grants were available, but they would not cover all the expenses of a three-year degree course. Coinciding with these problems was the fact that, in 1982, Lucy was in her final year at a junior state school, Newnham Croft. Stephen and Jane both wanted her to attend the Perse School, as her brother had done. With Timothy growing and everyday family expenditure increasing, there seemed to be no way for them to afford school fees for two children.

And what of the future? Stephen's illness had been stable for a number of years, but things could begin to slide again at any time – that was the nature of the disease. If he could no longer work, the prizes would soon dry up and his pension from the University could not sustain them comfortably. There was another great fear: if Jane could no longer look after Stephen and earn a salary, what would become of him? They did not like to discuss the awful possibilities, but they were there and had to be faced. They needed money, quickly. The last thing any of them wanted was for Stephen to end up in a nursing-home, if his condition should degenerate further, simply because they could not afford to look after him at home.

Something had to be done, and fast. Hawking had the germ of an idea in the back of his mind. He had mentioned it to no one, but had allowed it to grow and develop. Now, he realized, he would have to put his idea into action. It would be a number of years before Hawking's secret plan would come to fruition and, with one stroke, solve the family's financial problems. When it did, it was to change everything. But first there were intriguing developments to follow up in the field of inflationary cosmology.

13

When the Universe has Babies

Even though Hawking has offered us an image of a self-contained Universe, with no boundaries and no edges, either in space or time, many people still wonder what might lie 'outside' such a Universe. The analogy between the closed surface of the Universe and the closed surface of the Earth does, after all, encourage us to speculate that there might be other universes, just as there are other planets.

Within the framework of Hawking's no-boundary Universe, any such other worlds would have to be embedded in some strange form of space which has more than the three dimensions we are used to: the surface of a sphere, after all, is actually a *two*-dimensional surface wrapped around in the *third* dimension, but spacetime is four-dimensional; you always need at least one extra dimension to wrap up anything into a closed surface. But there is another model – or rather, series of models – developed from the inflationary scenario which offers us another way to imagine many worlds co-existing, without having to try to wrap our brains around the higher geometries of five or more dimensions (four of space plus one of time). Although Hawking himself has expressed reservations about the idea, which goes by the name of continual inflation, it is in fact based on his dramatic breakthrough discovery from 1974 that black holes explode.

Just after the Planck time, according to the inflationary scenario, the vacuum itself was in a 'false' state, excited and full of energy, like supercooled water. When the false vacuum underwent a transition into its stable, lower-energy state, this energy went into the phenomenal burst of expansion that is known as inflation, creating the smooth Big Bang out of which the Universe as we know it has evolved. But suppose this transition did not happen everywhere at the same time.

Almost as soon as Alan Guth came up with the idea of inflation, researchers such as Alex Starobinsky and Andrei Linde realized that different regions of the primordial false vacuum might have made the transition into the low-energy state independently. The effect would be rather like unscrewing the cap of a bottle of fizzy drink – a myriad of bubbles would appear throughout the fluid, each corresponding to a stable vacuum expanding in its own way. Unlike the bubbles in your fizzy drink, though, each of these bubbles would carry on expanding, until all the fluid had gone and only bubbles remained.

This possibility raised serious technical problems for early versions of the inflationary scenario, because if two or more expanding bubbles were to merge, they would create disturbances that would spread right through both bubbles. If we lived in a Universe that had formed in this way, it would not be perfectly uniform, because these disturbances would leave their mark – for example, on the microwave background radiation.

There are ways around this problem. The notion that Hawking himself favours is that of 'chaotic inflation', in which the world beyond our Universe (the infinite 'meta-universe') is in a messy state, with some regions expanding, some contracting, some hot and some cold. In such a chaotic meta-universe, there must inevitably be some regions just right for inflation to take place. We just happen, in this picture, to be in a Universe produced by a random fluctuation within the chaos.

But you don't have to invoke chaos to explain our existence.

Maybe we just happen to live in a bubble that hasn't (yet!) merged with any of its neighbours (if this sounds like an extraordinary coincidence, it may not be, as we shall see later in this chapter). Or perhaps some law of physics prevents bubbles from forming very close together in the 'fluid' of the false vacuum. This is where the proposal that Hawking Radiation might be involved comes in.

Hawking Radiation, as we saw in Chapter 9, is produced by the interplay of quantum effects and gravity at the horizon surrounding a black hole. But Hawking and his colleague Gary Gibbons, who shared an office with him in Cambridge in the late 1970s, realized that this kind of radiation must be produced wherever there is a horizon of this kind, and that such horizons do not always surround black holes.

Because of the way the Universe expands, the more widely separated two regions are, the faster they recede from each other. So regions of space that are far enough apart can never 'communicate' using light beams (or, indeed, anything else), because the space between them expands faster than light can travel. If light cannot travel from one region to another, then in effect there is a horizon which light cannot cross, separating the two regions of space as effectively as the horizon surrounding a black hole separates the inside from the outside.

Hawking and Gibbons showed that this kind of horizon will also produce radiation, just like the radiation at the horizon around a black hole, spreading out from the horizon into *both* regions of space. In the Universe as it is today, spread thin by expansion, the effect of this radiation is tiny, but it could have played a much bigger role in the early stages of the expanding Universe. The expansion of the Universe is steadily slowing down, as the gravity of all the matter in the Universe tries to pull everything back together in a Big Crunch. So the expansion rate was much faster, and the effect of Hawking Radiation from horizons therefore more pronounced, when the Universe was younger. Long ago, even rapidly separating regions had not had time to move far, and were much closer together.

The notion that radiation produced by horizons might affect the expansion of the Universe has been enthusiastically taken up, and combined with the idea of inflation, by Richard Gott of Princeton University. It has also been investigated by Andrei Linde, but he has made less noise about the idea than the ebullient Gott.

It turns out that under the right conditions the Hawking Radiation produced in a volume of space filled with horizons of this kind can provide the energy that drives inflation and makes the Universe (or rather, the meta-universe) expand superfast. The superfast expansion then creates more horizons, which in turn produce more radiation, driving the superfast expansion in a self-sustaining, continuing process of inflation. The bubbles of ordinary, low-energy stable vacuum that form within this infinite sea of inflationary expansion grow at a slower rate, and so even if two bubbles form next to each other they will be kept apart by the rapid growth of the false vacuum of the meta-universe between them.

The 'right' conditions for this process to work are mind-boggling. The temperature of the Hawking Radiation has to be about 10^{31} K, and the density of mass–energy in the false vacuum has to be an even more staggering 10^{93} grams per cubic centimetre. And everywhere throughout this extraordinary, rapidly expanding false vacuum, bubbles of stable vacuum are forming and becoming universes in their own right.

In this scenario, there is not just one Universe but an infinity of universes, forever separated from one another by the impenetrable walls of the superdense false vacuum. In a sense, such a concept is meaningless. The existence of other universes which we can never observe, and which can never have any interaction with our Universe, is a matter more suitable for discussion among philosophers than astrophysicists. But it turns out that there are more ways than one to make a universe, and that in some scenarios universes *can* interact with one another, producing consequences of interest to everybody, not just to astrophysicists and philosophers.

With all this talk of superdensity and superenergy, and numbers like 10^{93} grams per cubic centimetre being bandied about, it is natural to wonder how much mass–energy our entire bubble Universe contains, (assuming, that is, that any of these scenarios have a grain of truth in them). The answer is perhaps even more startling – none at all! Let us leave the discussion of continual inflation to the philosophers, and look again at Hawking's no-boundary model of the Universe to see how this can possibly be true.

We are used to thinking of mass–energy chiefly in terms of lumps of matter: stars, planets, and so on. Each of them contributes its own amount of mc^2 to the total mass–energy of the Universe. But there is another, equally important contribution (*exactly* equally important, if Hawking's ideas are correct). It comes from gravity. And there is a strange thing about gravitational energy – it is negative.

To understand what this means, physicists talk in terms of the gravitational energy of a hypothetical collection of particles. This is zero if the particles are dispersed to infinity, spread apart from one another as far as possible. But if the collection of particles falls together under the influence of gravity, perhaps eventually to make a star, it *loses* gravitational energy. Since the particles start with zero energy, this means that by the time they have collected together to form a star or a planet they have negative energy. And if all the matter in the entire Universe could be collected together at a single point, its negative gravitational energy ($-mc^2$) would exactly cancel out all the positive mass–energy ($+mc^2$) of all the matter.

But that is exactly how we think the Universe did start out: with all its mass–energy concentrated in a point. The closed Universe scenarios actually describe a situation in which a point of zero energy becomes separated into matter with positive energy and gravity with negative energy, expands out to a certain size, and then collapses back into a point of zero energy again. At first the idea seems ridiculous. However, this is not some crackpot, lunatic-fringe theory, but a respectable cosmological idea, backed up by the equations of relativity.

The Universe, it seems, is the ultimate free lunch. And if the Universe contains zero energy, how much energy does it take to make a universe? Not a lot – certainly not very much compared with the amount of mc^2 contained in your body, or the pages of this book. For, according to Alan Guth and his colleague Edward Fahri, all you need is enough energy to squeeze some matter into forming a black hole. Then, the new universe comes free – one universe free with every black hole. In a *tour de force* to rank with the great conjuring tricks, Guth and Fahri have shown that the two great threads of Hawking's life's work are really one and the same: black holes *are* big bangs.

In principle, the seeds of entire universes could be produced out of nothing at all, in a manner reminiscent of the way pairs of virtual particles can be produced out of nothing at all by quantum uncertainty (as we saw in Chapter 9). Such a baby universe would be in the form of a superdense concentration of mass, smaller than a proton but containing no energy because the mass is balanced by negative gravitational energy. Of course, according to the ideas of the 1970s and before, such tiny superdense seeds would immediately collapse back into nothing under their own weight. But inflation provides a way to blast out such a seed to form an expanding universe before gravity can make it collapse. It would then take many billions of years for gravity first to halt the expansion and finally to make the universe disappear into a Big Crunch.

So do we really need the continually inflating false vacuum to make bubble universes pop up in infinite numbers? At first sight, this raises a worrying possibility. If a bubble universe can pop into existence out of the ordinary vacuum, what would happen if one burst into existence near us? Would we be overwhelmed by the expanding fireball of a Big Bang going on right next door? Fahri and Guth think that there is nothing to worry about. If such baby universes pop into existence spontaneously, or if they were created artificially, they would have no further interaction with our Universe once they had been born.

Remember that the seed of such a bubble universe must be self-contained, destined ultimately to collapse back in on itself; in other words, it must be a black hole. Fahri and Guth found that you could trigger this process of universe creation artificially, by squeezing a small amount of matter into a black hole at a temperature of about 10^{24} K (quite modest compared with conditions in the false vacuum). But they gave their scientific paper on the subject the tongue-in-cheek title 'An Obstacle to Creating a Universe in the Laboratory',[1] pointing out that although we have the technology (hydrogen bombs) to do half the job, releasing the energy required, we don't yet have the ability to confine the energy released by hydrogen bombs within a black hole.

But it is not beyond the bounds of possibility that a civilization more advanced than our own might be able to confine the required energy in a small enough volume. What would happen then? To the people who created this energetic minihole, very little. The black hole would simply form, spend billions of years evaporating through Hawking Radiation, and then disappear. But within the horizon of the hole things would be very different.

According to the calculations by the American team,

conditions inside such an energetic minihole will sometimes be such as to trigger inflation. But when such a baby universe begins to expand, it does so not by bursting out of the minihole to engulf its surroundings in the spacetime in which it was created, but by expanding in a set of directions which are *all* at right angles to *each* of the dimensions of the parent universe. And exactly the same thing will happen to baby universes that are produced by natural quantum fluctuations.

Because all the sets of dimensions are at right angles, the different universes never interact with one another once they have formed. But there is a crucial difference with the continual inflation idea, where the bubbles never interact at all. In the scenario sketched by Fahri and Guth (and studied by others, including Linde), one universe is created *from* another. In this picture, our Universe is the progeny of a previous universe; and it is even possible that our expanding bubble of spacetime was created artificially in the equivalent of a laboratory in that parent universe. Science-fiction writer David Brin is already working on the implications in a linked series of stories; we will leave further speculations along these lines to Brin and his colleagues, while we try to explain the implications in terms of the spontaneous creation of baby universes.

It is hard to get a mental grip on the proliferation of dimensions that this implies. Every baby universe will contain its own vacuum, within which other quantum fluctuations can occur, producing yet more baby universes each with their own set of dimensions, with every set of dimensions at right angles to every other set. As usual, we have to fall back on an analogy in two dimensions, bent round a third, to get a picture of what is going on.

The helpful image is the old, familiar one of the Universe represented by the skin of an expanding balloon. What we have to imagine now is that a tiny piece of that skin is pinched off, forming a little blister connected to the Universe by a narrow throat – the black hole. That little blister can now, in turn, expand to enormous size, while all that any resident of the parent universe sees is the tiny black hole throat in the fabric of spacetime. And the whole process can repeat indefinitely, producing an infinite foam of bubbles, each one a universe in its own right. Quantum cosmology actually allows the possibility of creating not just one Universe, but an infinite number of universes, out of nothing at all.

And this raises another question. At one level, physics operates by finding out the rules according to which the Universe operates, and using them to make predictions about how systems will interact. We find, for example, that the speed of light has a certain value, and that this is the ultimate speed limit. That enables us (or, at least, it enabled Einstein) to work out how our view of the world changes when we move at high speed. But at another level, some physicists puzzle over *why* the rules should have the precise form that we find.

Why, for example, is the speed of light 300,000 kilometres a second, rather than, say, 250,000 kilometres a second? Why does Planck's constant have the precise value it has, and not one a little bigger or a little smaller? What would happen if gravity were weaker (or stronger)? And so on. We live in a world that seems to be just right for life-forms like us – which is in a way tautological, since obviously if the world were very different we would not be here to wonder about these things. But as far as anyone is yet able to tell, the rules of physics that came out of the era of inflation could have been different from the rules we know, either subtly different, or dramatically different. Is it, then, just a coincidence that these rules have produced a Universe suitable for people like us to live in? The idea of an infinity of bubble universes, either formed out of an eternally expanding false vacuum or pinched off from one another by the baby process, says that it is not – and explains other cosmic coincidences as well.

*

The idea of trying to understand the nature of the Universe in terms of the relationship between the laws of physics and ourselves is known as 'anthropic cosmology'. It has a long history, but in its modern version it stems mainly from a revival of interest triggered by Martin Rees, of the University of Cambridge, in the 1970s, and continuing to the present day.

Rees is an exact contemporary of Hawking. He was born on 23 June 1942, when Hawking was six months old. They were working for their PhDs in Cambridge at the same time, and Rees became Plumian Professor of Astronomy and Experimental Philosophy in 1973, at the remarkably early age of thirty-one, just six years before Hawking became Lucasian Professor. He was elected a Fellow of The Royal Society in 1979, five years after Hawking. But where Hawking has made his reputation by investigating in great detail one particular set of problems – the singularities and horizons around black holes and at the beginning of time – Rees is known and respected for the breadth of his work, ranging from quasars and pulsars to the influence of black holes on their surroundings, cosmology, and the nature of the dark matter that holds the Universe closed. When he turned his attention to anthropic cosmology, and stirred the revival of serious interest in the subject by scientists in the 1970s and 1980s, for once Hawking was prepared to follow somebody else's lead.

Rees has developed a particularly nice example of the nature of anthropic reasoning in cosmology. He has worked out in detail the evolution of a universe in which gravity is stronger than in our Universe, but every other rule of physics is the same. Galaxies, stars and planets can all exist in this model universe, but they are all very different from their counterparts in our Universe. In particular, everything is speeded up to such an extent that it is doubtful whether intelligence (which has taken more than four billion years to emerge on Earth) could ever evolve.

For the particular value of the strength of gravity chosen by Rees, each star has a mass about the same as that of an asteroid in our Solar System (much less than the mass of the Moon) and a diameter of about two kilometres. The typical lifetime of such a star is just one of our years, and it burns with a brightness one hundred-thousandth that of our Sun. The Earth has an average surface temperature of about 15 °C, and a planet in this other universe, orbiting around its parent star at a distance roughly twice as great as the distance from the Earth to the Moon, would have a similar surface temperature. It would take about twenty of our days for the planet to orbit the star. So, with the star itself having such a short lifetime, it would be burnt out in just about 15 of the planet's 'years', whereas the lifetime of our Sun is likely to be at least 10 billion of our years.

Life on the surface of such a planet would be short, in more ways than one. The biggest mountains on the tiny planet could be no more than 30 centimetres high, while the maximum mass of any creatures roaming its surface would be just one-thousandth of a gram – any bigger than this, and their bodies would break if they fell over in the strong gravity of that world.

And all of these dramatic changes stem, remember, from making a change in just *one* of the constants of physics, the strength of gravity! It is possible to imagine very many changes that would ensure that the universe that emerged from the inflationary phase would be quite inhospitable for life-forms like us.

If ours is the only possible Universe, then the existence of the cosmic coincidences that permit our existence is a real puzzle. But if there are many possible universes, then there is a straightforward explanation. Every different bubble universe may have its own laws of physics. In some cases, that will mean that the bubbles are held together very tightly by gravity, and recollapse before life can evolve. In others, gravity may be so weak that material is never pulled together to form stars and planets at all. But there will be a range of possibilities – a range of universes – where stars, planets and intelligences can evolve. The same argument applies to each and every one of the exact values of the laws and constants of physics.

If this picture is correct, it means that there may be an infinite number of universes in the meta-universe, and out of that infinite number life-forms like ourselves will exist only in universes where the laws of physics are just right. The fact that we exist pre-selects, to some degree, the exact rules of physics which we will discover the Universe operates on. This idea is known, rather grandly, as the 'anthropic principle', a term coined by Bernard Carr, who worked with Rees on a seminal paper on the topic.

Of course, because the different universes can never communicate with one another, this is still largely a matter for the philosophers to debate. Except for one thing. Remember that the crucial ingredient of Hawking's no-boundary model is the sum-over-histories quantum approach. When we mentioned this earlier, we rather glossed over the explanation of what, exactly, the different histories that were being 'summed' were. Now we can set the record straight.

Instead of regarding all the different possible universes that could have emerged out of inflation, each with its own set of physical laws, as 'real', we can regard them as mathematical possibilities, like the many different paths that an electron can take from A to B. And, using the sum-over-histories approach, Hawking shows not only that our Universe is one of the *possible* histories, but also that it is one of the *most probable* ones:

> . . . if all the histories are possible, then so long as we exist in one of the histories, we may use the anthropic principle to explain why the universe is found to be the way it is. Exactly what meaning can

be attached to the other histories, in which we do not exist, is not clear.[2]

Nevertheless, using the 'no-boundary' condition, Hawking and his colleagues have found that the Universe must start out with the maximum amount of irregularity allowed by quantum uncertainty, and that inflation and the subsequent more leisurely expansion of the Universe then make these irregularities grow to become the clouds of gas that then contract to become galaxies of stars within the expanding Universe.

All of this is very much research at the cutting edge of science today. The choice of different variations on the theme – bubbles in a continually inflating false vacuum, baby universes, a choice of quantum histories – reflects not an inability of physicists to make up their minds, but an attempt to push ahead on many different fronts, not yet knowing which (if any) will turn out to hold the most promise in the long term. But it is already clear that in the 1990s the basic premisses underlying cosmological thinking have changed dramatically from those of what we might call the 'pre-Hawking' era. Thirty years ago, it was generally accepted that our Universe was unique. Today, it seems to be generally accepted that, one way or another, it is just one among many. Is it any wonder that, when Hawking presented these ideas in a book in 1988, the book took the world by storm?

14

A Brief History of Time

The dying notes of Tears For Fears' 'Mad World' lead into Radio One's 12.30 p.m. news as Simon Mitton walks into the DAMTP and a car with its window down and radio turned up parks in front of the building on the other side of the cobbled courtyard. The news report is full of peace protesters at Greenham Common, British troops in the troubled city of Beirut, and the biggest Christmas film ever, *E.T.*, but Mitton has other astronomical thoughts on his mind. He is visiting Stephen Hawking to discuss the imminent publication of the professor's new book for Cambridge University Press, *The Very Early Universe.* Unexpectedly, however, after talking through the latest details of the book over tea and biscuits, the two of them fall into a discussion about something altogether different – a popular cosmology book, which Stephen has been mulling over for some time.

For almost as long as he had known him, Mitton had been intimating to Hawking that he should attempt a cosmology book aimed at the popular market. Hawking had displayed little interest in the idea, but by late 1982 he had come to recognize that such a project might provide the answer to his looming financial difficulties, and he decided to revive the idea. The two of them had enjoyed a fruitful publishing relationship for many years, and, despite the problems over *Superspace and Supergravity*, Hawking's first thought was to approach Cambridge University Press with the proposal. Mitton's original intention was for Hawking to attempt a book on the origin and evolution of the Universe. Cambridge University Press had enjoyed a long tradition of publishing popular science books written by eminent scientists, such as Arthur Eddington and Fred Hoyle, whose titles had sold well. A popular book by Stephen Hawking would, he believed, neatly follow on from these.

According to Mitton, Hawking laid things on the line immediately. He wanted a lot of money for this book. Mitton had always known him to be a tough negotiator; that was clear from the fracas over the cover for *Superspace and Supergravity*. When it came to financial matters he was prepared for some intransigence, but in the event even Mitton was surprised by Hawking's suggestions. At their first organized meeting to discuss the book, Hawking opened the conversation by explaining his financial situation, making it clear that he wanted to earn enough money to continue financing Lucy's education and to offset the costs of nursing. He was obviously unable to provide any form of life insurance to protect the family in the event of his death or complete incapacitation, so if he was going to spend a considerable amount of his valuable time away from research writing a popular book, he expected an appropriate reward.

Mitton is philosophical about the whole matter, pointing out that Hawking was showing remarkable loyalty towards Cambridge University by staying there. There is absolutely no doubt that he could have commanded a huge salary from any university in the world. A number of colleges in the USA would have offered him six-figure sums simply for the prestige accompanying his international fame, not to mention the enormous kudos of cashing in on the important breakthroughs he would almost certainly make in the near future. The fact that he remained in Cambridge for a fraction of the salary he could command elsewhere is, Mitton believes, a great credit to him. The simple fact is that the Hawkings loved Cambridge. They had lived there for nearly two decades, and Stephen had spent practically all of his academic life at the University. The DAMTP is, without doubt, one of the best theoretical physics departments in the world, and he would have left it only as a last resort.

In the early eighties, Simon Mitton's office was based in the same courtyard as the DAMTP in Silver Street, so the two of them had plenty of opportunity to talk about the proposed project. One afternoon, Hawking went to see him with the rough draft of a section of the proposed book. Mitton knew the commercial market as well as any publisher. In fact he was by that time author of several successful popular science books himself. He had a very clear idea of the type of book the general public would want and which would earn Hawking the sort of money he was after. After looking through the section Hawking had shown him, he came to the conclusion that it was far too technical and highbrow for the general reader. 'It's like baked beans,' he told Hawking. 'The blander the flavour, the broader the market. There simply isn't a commercial niche for specialist books like this, Stephen.'

Hawking went away and thought about Mitton's comments; Mitton went to the Cambridge University Press Syndicate to see what they thought of the idea. The two men met up again shortly afterwards. Mitton had the encouraging news that the Syndicate had accepted the idea of the book with glee, and had handed over to him all negotiations. Hawking, for his part, had done a little editing of the section he had written earlier. Mitton sat back and flicked through the manuscript as Hawking remained motionless in his wheelchair on the other side of the room, patiently awaiting his opinion. Finally, Mitton put the typescript down on his desk and looked across at him.

'It's still far too technical, Stephen,' he said at last. Then smiling, he made the now famous statement: 'Look at it this way, Steve – every equation will halve your sales.'

Hawking looked surprised. Then, smiling, he said, 'Why do you say that?'

'Well,' replied Mitton, 'when people look at a book in a shop, they just flick through it to decide if they want to read it. You've got equations on practically every page. When they look at this, they'll say, "This book's got sums in it," and put it back on the shelf.'

Hawking took Mitton's point. Over a cup of tea, the two of them began to talk money. Mitton suggested an advance, to which Hawking smiled and made a faintly disparaging reply. Mitton knew this was going to be tough. By the end of the afternoon, Hawking had talked Mitton into a £10,000 advance, by far the biggest Cambridge University Press had ever offered anyone. The percentage royalties on both the hardback and the paperback were also excellent. The next morning, Mitton sent a contract over to Hawking's office. He never heard from him on the matter again.

Early in 1983, as Stephen Hawking and Simon Mitton sat in an office in Silver Street, Cambridge, discussing over tea the idea of doing a popular book, three thousand miles away a tall, bearded man in his early thirties passed by a news-stand on Fifth Avenue. Stopping briefly to scan the titles, he picked up a copy of the *New York Times*, paid for it and walked on. Arriving at his office a few blocks away, he sat down at his desk; he had a few moments to spare before his lunch appointment with a literary agent at a local restaurant. Peter Guzzardi opened the paper and the magazine fell out on to the desk. There, on the front cover, a picture of a man in a wheelchair was staring back at him. Discarding the rest of the newspaper, Guzzardi quickly turned to the cover article, 'The Universe and Dr Hawking', and began to read.

Within minutes he was hooked. The article described the amazing story of the crippled Cambridge scientist, Stephen Hawking, who had revolutionized cosmology and had, for the past twenty years, successfully overcome the devastating symptoms of a wasting neurological disease called amyotrophic lateral sclerosis. By the time he had finished the article, he knew he had stumbled upon a great story, and being a senior editor at Bantam Books he was in a perfect position to do something about it. With the enormous possibilities opened up by his discovery already racing around in his mind, he stuffed the magazine into his bag and headed off for lunch.

Peter Guzzardi's appointment was with the agent Al Zuckerman, who was president of a large agency called Writer's House based in New York City. Over dessert, Guzzardi mentioned what he had just been reading about Stephen Hawking. Zuckerman had read the same article, and was already on the case. He had recently heard through a mutual friend, a physics professor at the Massachusetts Institute of Technology called Daniel Freedman, that Hawking was working on a book. He had then contacted his own brother-in-law, who just happened to be a physicist himself, about the project.

By the time of his lunch engagement with Peter Guzzardi, Zuckerman had already decided to get in touch with Hawking to establish the state of play. Seeing the human story behind the discoveries, and believing that such a book could be hugely successful, he wanted to be involved. Before they left the restaurant, Peter Guzzardi made it very clear that if Zuckerman were to meet Hawking and discover that he had not already signed to another publisher, he was sure Bantam would like to know about it. On the sidewalk, the two men shook hands and went back to their respective offices.

Six months passed before Peter Guzzardi heard another thing from Al Zuckerman, but the agent had not been idle in the meantime. He had succeeded in contacting Hawking on the point of signing a contract with Cambridge University Press (CUP). He had cut it fine – a few days later the deal would have gone through, and Zuckerman would have had little incentive to get involved. Although CUP would undoubtedly have made a very good job of producing Hawking's book, they probably would not have been the right publishers for it. Hawking wanted to sell his book in vast numbers, tapping the popular market; as a highly prestigious academic publisher, CUP are simply not geared up for that area of the business.

Dennis Sciama recalls how he met Hawking on a train around the time of CUP's offer, and discovered that his former student was working on a popular book.

'You're doing it with CUP?' he asked.

'Oh no,' Hawking replied with a mischievous grin. 'I want to make some money with this one.'

Zuckerman managed to persuade Hawking not to sign the contract before giving him a chance to see what he could do. They agreed that if he was unsuccessful in placing the book, then Hawking could always fall back on the offer from CUP, but Zuckerman had a very strong feeling that he could get more than £10,000 in advance and hook one of the big trade publishers. Hawking drafted a proposal for the book and produced a sample section of around a hundred pages, and Zuckerman contacted a number of publishers, including Bantam, in the States. He had decided at the beginning that he would go for a deal with an American publishing company first, and secure contracts for publication in other countries at a later date.

Peter Guzzardi received the proposal early in 1984 and presented it to the next scheduled editorial meeting. He took with him the *New York Times* article which had attracted his interest in the first place, and showed it to his colleagues. They immediately saw the potential of the proposal, and needed little persuading that the idea was a good one. By the end of the meeting they had agreed to make a serious bid for the rights.

Despite Bantam's obvious interest, Zuckerman decided to hold an auction for the book. The whole thing was conducted over the phone. He sent out the package Hawking had put together to a collection of major publishers and told them that if they were interested in the book they had to make an offer by a certain pre-arranged date. Interested parties then contacted him with their offers and were told if there was a rival offer which bettered their own. They then had the choice of upping their offer or dropping out of the bidding. Towards the end of the auction day, two rivalling publishers were left to compete over the contract: Norton and Bantam. Norton had recently published *Surely You're Joking, Mr Feynman!*, the autobiographical reveries of the Nobel Prize-winning Caltech professor, Richard Feynman, and were very keen on the Hawking proposal. The Feynman book had done exceptionally well. To them the market potential of a popular book by Hawking was obvious.

As evening approached and the two companies upped their bids further, Bantam decided to take the plunge and make a final, some may have thought hugely risky, offer. Hurried telephone calls flashed between offices and hastily arranged meetings were held to decide what should be done. Finally, Guzzardi was given the go-ahead to make his final bid. He offered a $250,000 advance for the United States and Canada and a very favourable deal on hardback and paperback royalties. As the sun set over the ragged skyline, tense minutes turned into a nail-biting half-hour in Guzzardi's office in Manhattan. He really wanted this book.

Finally the phone rang. Guzzardi grabbed for it. Norton had not matched the offer. Subject to Hawking's approval of a submitted letter outlining what they would do in terms of re-writes and promotional technique, the book was theirs.

Author and agent obviously had little doubt of the book's worth and the saleability of the Hawking name – remarkably cool behaviour, on Hawking's part, for a man who, for all his fame and earning potential abroad, was in reality in a rather delicate financial state. Peter Guzzardi accepted the conditions and wrote to Hawking with his ideas. Hawking obviously approved, for the contract was signed a short time later. Guzzardi says that one of the things he believes clinched the deal was his suggestion that the book should be on sale at every airport in America. Hawking loved the idea. The fact that his book was with one of the world's biggest publishers gave him a real thrill.

Guzzardi first met Hawking after a conference at Fermilab, the high-energy physics research establishment just outside Chicago. He remembers that Hawking was very tired after delivering his talk, but was still very receptive and enthusiastic about the project. Recalling his first impressions of Hawking, he said, 'The man has a formidable presence. He came across as a very powerful personality.'

By this time, Hawking's lectures were always delivered via an interpreter, usually a research assistant who would handle the slide projector and present Hawking's pre-scripted lecture. The same interpreter acted for Guzzardi when he met Hawking after the talk. 'It was a bit like listening to someone speaking in a foreign language,' Guzzardi recalls. 'You pick up a sort of rhythm, without actually understanding what he's saying.'

Although Hawking was very happy to discuss the book even at the end of a tiring day, Guzzardi sensed that some of his assistants were less than enthusiastic about the whole thing. He feels they resented the idea that their professor's work was being popularized, and that attempts to reinterpret his theories in layman's language for public consumption would somehow devalue them. According to Guzzardi, this was an attitude Hawking did not share with his students; on the contrary, he was very much in favour of communicating his theories to a general audience. Before *A Brief History of Time*, Hawking had shown great interest in delivering public lectures about his work and had, Guzzardi feels, a definite sense of mission about public awareness of cosmology.

After the first meeting, an exchange of letters between Cambridge and New York began in which suggestions and counter-suggestions were made about passages in the growing manuscript. Throughout the long gestation period of the project, Guzzardi sought advice from other scientists and expert communicators to help him understand Hawking's ideas, feeding back his digested version of their remarks to steer Hawking further in the direction in which he had said he wanted to go – towards a best-seller. Considering Hawking's commitments to the DAMTP, his busy schedule of talks and lectures around the world, and his family responsibilities, work on the book progressed well. But despite all their efforts it was to take a further eighteen months before Hawking and Guzzardi succeeded in knocking the manuscript into shape and preparing it for publication.

There have been suggestions that at one stage Bantam wanted the book ghost-written by a successful science writer but that Hawking totally rejected the idea. Such notions are completely unfounded. At no time was such a suggestion made by Guzzardi. In fact, it had been Al Zuckerman who had initially proposed the idea:

> I read the manuscript and thought it was very interesting, and that I could certainly find a publisher for it, but that it would not be readily comprehensible to the lay reader . . . I thought at that time that we should bring in a professional writer to help put the ideas across in language which would be more easily understood. Hawking refused; he wanted the book to be all his. And he is a very strong-minded man.[1]

In his role as editor, Guzzardi tried to put himself in the position of the average man in the street buying and attempting to read the book. He tried to convey this to Hawking during their transatlantic correspondence, with remarks like

'I'm sorry Professor Hawking, I just don't understand this!' Zuckerman has said of Guzzardi's efforts:

I would guess that, for every page of text Peter wrote two to three pages of editorial letters, all in an attempt to get Hawking to elaborate on ideas that his mind jumps over, but other people wouldn't understand.[2]

'I was persistent,' Guzzardi says, 'and kept on until Hawking made me understand things. He may have thought I was a little thick, but I risked it and kept on plugging away until I saw what he was talking about.' According to Guzzardi, Hawking was perfectly amiable about the whole thing and showed great patience with him. In his typically modest way, he also claims that Hawking gave him too much credit in the book's acknowledgements. 'I did,' he points out, 'what any normal intelligent person would do and persevered until I understood what was going on.'

Kitty Ferguson, in her book *Stephen Hawking: A Quest for the Theory of Everything*, has suggested that because of his condition Hawking's use of few words in his explanations meant that in lectures and seminars he would often jump from thought to thought, wrongly assuming that others could see the connection. Without careful editing, this could obviously present serious problems in a supposedly popular science book.

For Peter Guzzardi, the responsibility of editing *A Brief History of Time* was a very exciting experience. He realized before the contract was even signed that Hawking was the man to write the definitive work on the theory of the origin and evolution of the Universe. It was he who had done the seminal work on many of the new ideas at the heart of the subject. Who could have been more suited to the task? The manuscript was from the horse's mouth. Guzzardi is from the school of thought that proposes Hawking as the Einstein of the latter half of the twentieth century. Although he is not himself a scientist, through their collaboration on the book he undoubtedly grew to know Hawking and his way of thinking very well. His understanding of the man is very different to the way his students and professional associates understand him, but perhaps equal in depth.

To many, Hawking is not the hero the public seem to have made him. There are those who suggest that he is melodramatic at conferences, that he is pretentious and showy, that his constant questioning is affected and deliberately argumentative.

The physicist and popular writer Paul Davies has pointed out that there can be few things more intimidating than for Hawking to come crashing through the doors of a lecture theatre five minutes after an inexperienced speaker has begun to talk. Even worse are the occasions when he decides to leave before the end of a lecture and goes careering along the aisle, accelerating his motorized wheelchair straight towards the swing-doors at the back of the room. But Davies admits:

> Often, it is simply that Stephen is hungry or has remembered that he must phone someone urgently. His lateness is always unintentional and not done to intimidate, but fortunately it hasn't happened to me – yet!

There are those who do not view Hawking's antics and celebrity so kindly. One theorist has been quoted as saying, 'He's working on the same things everybody else is. He just receives a lot of attention because of his condition.'[3]

Do Hawking's critics have a point, or are such statements simply sour grapes over the hype surrounding him? Hawking's own opinion on people comparing him with Einstein is typically brash: 'You shouldn't believe everything you read,' he says with an ambiguous smile.[4]

Finishing the first draft took up most of 1984. It was the year a bomb planted in the Grand Hotel in Brighton nearly killed the British Cabinet, and the Prime Minister of India,

Indira Gandhi, was assassinated by her own bodyguards in the garden of her New Delhi home.

As the months passed and Hawking juggled his commitments, the manuscript grew and the stack of correspondence with his editor expanded apace. In the world at large, a baboon's heart was transplanted into a fifteen-day-old baby, Bishop Tutu received the Nobel Peace Prize, and towards the close of the year Ronald Reagan was re-elected as US president.

The first draft of the manuscript was completed by Christmas, and work on re-writes began in the New Year. The exchange of letters between Hawking in Cambridge and Guzzardi in Manhattan became even more frenetic as the deadline approached.

The trade press got wind of the book soon after Christmas 1984, but appeared to be nonplussed by seemingly misplaced enthusiasm at Bantam:

Is it the imminence of spring, or is the new enthusiasm we detect genuine? Everywhere we hear the sound of feet jumping up and down in sheer elation over some pet project. At Bantam, Peter Guzzardi is jumping for joy over the acquisition of Stephen Hawking's *From the Big Bang to Black Holes* . . . Paying what Guzzardi calls 'significant six figures, definitely above $100,000,' Bantam has plans to publish the book in hardcover 'sometime in 1986' . . . 'It's a great book to have,' enthuses Peter. 'Hawking is on the cutting edge of what we know about the cosmos. This whole business of the unified field theory, the conjunction of relativity with quantum mechanics, is comparable to the search for the Holy Grail.'[5]

The mid-eighties was indeed a time of growing optimism. As the major nations of the world dragged themselves out of recession, markets began to expand and all sectors of business were on the up. It was the era of the yuppie. The 'city-slicker' emerged metamorphosed from the post-hippy hibernation of the seventies, cast off the clinging remnants of introspection and integrity, and jumped into a Porsche 911 convertible.

Newly elected right-wing governments were in power in the major industrialized nations and you could almost smell the odour of growing confidence in the spring air. Life was good; no one had noticed the swelling bass note of over-expansion and downturn. Share prices in champagne and designer labels rocketed, and big publishing deals became part of the norm.

In July 1985, Hawking decided to spend some time at CERN, the European organization for nuclear research in Geneva. There he could continue with his fundamental research and also allow himself time to devote to what he described to friends as 'a popular book'. He rented an apartment in the city where he was looked after by a full-time nurse and his research assistant at the time, a French Canadian called Raymond Laflamme. Jane, in the meantime, had decided to tour Germany to visit friends. The couple planned to meet up in Bayreuth to attend the Wagner Festival in August, after Stephen had completed the re-writes for the book.

One evening at the beginning of August, Hawking retired to bed late after a long day making corrections to the manuscript. His nurse helped him into bed, and sat down to relax in an adjoining room. After finishing a magazine article, she would begin her routine of checking her patient every half-hour throughout the night. Around 3 a.m the nurse walked into Hawking's room to find him awake and having problems breathing. His face had turned violet and he was making a gurgling sound in his throat. She immediately alerted Laflamme, and an ambulance was called.

Hawking was rushed to the Cantonal Hospital in Geneva, where he was immediately put onto a ventilator. Legend has it that it was thanks to television that the doctor in charge of receiving the crippled scientist at the hospital saved Hawking's life. Shortly before Hawking had become one of his patients, he just happened to watch a TV programme about a

Cambridge physicist who suffered from ALS. Knowing Hawking's condition, he knew which drugs he could and could not give to his patient. A doctor who had not been fortunate enough to have caught the programme may well have killed him unintentionally.

Hawking was rushed to intensive care, and the authorities at CERN were notified. The Division Leader, Dr Maurice Jacob, arrived at the hospital before dawn and was informed that things were touch and go. It was thought that Hawking had suffered a blockage in his windpipe and was suspected of having contracted pneumonia. ALS sufferers are more susceptible to the disease than others; in many cases, it proves to be fatal. Maurice Jacob and his staff immediately tried to contact Jane, but it proved no easy matter. She was travelling from city to city and had left a series of telephone numbers with Stephen's nurse. The problem was that no one was absolutely sure of her schedule. Frantic calls were made to various private numbers in Germany, until she was finally tracked down at a friend's house near Bonn.

Jane arrived at the Cantonal Hospital to find her husband in a very bad way. He was on a life-support machine, but out of immediate danger. However, in the opinion of the doctors, he would have little hope of survival without a tracheostomy operation. Stephen was unable to breathe through his mouth or nose, and would suffocate if he was taken off the ventilator that stood beside his hospital bed. The operation involved slicing in to the windpipe and implanting a breathing device in his neck, just above collar-level. Jane was told that the operation was essential to save her husband's life, but there was a major snag. If they went ahead, he would never be able to speak or make any vocal sound again.

What was she to do? The decision could come only from her. Although Stephen had hardly been capable of speech for many years, with only his family and close friends able to understand him, there was now the prospect of total loss of communication. His voice may have been difficult to understand, but it was still speech. There was, she knew, a technique for recovering some speech after a tracheostomy, but that was a possibility only if the patient was reasonably fit. The doctors around her were staggered that a man in Hawking's state could still be travelling the world, but there would be no chance of his regaining any form of speech in his physical condition. Could she take the decision to go ahead with it and condemn her husband to silence?

> The future looked very, very bleak. We didn't know how we were going to be able to survive – or if he was going to survive. It was my decision for him to have a tracheostomy. But I have sometimes thought, what have I done? What sort of life have I let him in for?[6]

After the operation Hawking remained in the Swiss hospital for another two weeks. An air-ambulance then returned him to Cambridge, where he was admitted to Addenbrooke's Hospital. The plane flew in to Marshall's Airport, where he was met by doctors and escorted to the intensive-care unit at the hospital.

That evening the senior nurse of the medical unit at Addenbrooke's was quoted in the *Cambridge Evening News* as saying, 'He is going into intensive care. We are not sure of his condition and he needs to be assessed.'[7] In the event, he was to spend a further few weeks in hospital in Cambridge before he was allowed home to West Road.

In many respects, Hawking had been lucky once again. He had survived by the skin of his teeth. Many ALS sufferers die from pneumonia initiated by their condition. When he caught the infection he just happened to be in one of the most medically advanced countries in the world; he was received at the hospital by a doctor who had recently seen him on TV and knew of his condition; and he had the support of an intelligent, caring wife. However, one of the most serendipitous facts of all is that, if he had contracted pneumonia two years earlier, things would have been far worse.

In August 1985 the writing of what would become the best-selling *A Brief History of Time* was almost complete. Peter Guzzardi had, of course, been notified immediately Stephen had fallen ill and had continued editing the manuscript while Hawking was recovering in hospital. The family had received some money from the advance and could just about cope financially with the immediate crisis. The problem for Jane, however, was what would happen in the long term. After the tracheostomy, Stephen would need round-the-clock nursing. The best the National Health Service could offer was seven hours' nursing help a week in the Hawkings' home, plus two hours' help with bathing. They would have to pay for private nursing. The advance from the book would not last long, and there was absolutely no certainty about its eventual success. To Jane, there seemed little long-term hope. How were they to survive if he could never work again?

There were few possibilities. She would willingly have left her own career and devoted herself full-time to looking after her husband, but she was not a qualified nurse, and in any case, who would then provide for the family? The alternative was the dreaded thought of Stephen in a nursing-home, unable to work, slipping into gradual decline and eventual death. 'There were days when I felt sometimes I could not go on because I didn't know how to cope,'[8] Jane has said of the period.

It was obvious they would have to find financial support from somewhere. Jane wrote letter after letter to charitable organizations around the world, and called upon the help of family friends in approaching institutions that might be interested in assisting them. Help arrived from an American foundation, aware of Hawking's work and international reputation, that agreed to pay £50,000 a year towards the costs of nursing. Shortly afterwards several other charitable organizations on both sides of the Atlantic followed suit with smaller donations. Jane feels bitter about the whole affair. She resents the fact that, after paying a lifetime of contributions to the National Health Service, they were offered such meagre help when the need arose. She is very aware that if her husband had been an unknown physics teacher, he would now be living out his final days in a residential home. 'Think of the waste of talent,'[9] she has said of the situation.

The very month in which the Hawkings received the offer of financial support, a computer expert living in California, Walt Woltosz, sent Stephen a program he had written called Equalizer. It was compatible with the computers he used at home and in the office, and enabled him to select words on a screen from a menu of 3,000. He could move from word to word by squeezing a switch held in his hand. Tiny movements of his fingers were enough to operate the device and move a cursor to the desired word. When a sentence had been built up, it could be sent to a voice-synthesizer that then spoke for him. Certain key sentences were pre-programmed into the computer to speed up the process, and with a little practice Hawking found he could manage about ten words a minute. 'It was a bit slow,' he has said, 'but then I think slowly, so it suited me quite well.'[10]

Hawking's new computer-generated voice completely transformed his life. He could now communicate better than he could before the operation, and he no longer needed the help of an interpreter when lecturing or simply conversing with people. Since the tracheostomy, his only means of communication had been by blinking his eyes, spelling out words written on a card held in front of him. The voice-synthesizer has a definite accent, variously described as American or Scandinavian. Because there is a degree of intonation on certain words, it doesn't sound too much like a robot –

something Hawking would have hated. He really wishes that the synthesizer could produce a British accent, and he often greets people with, 'Hello, please excuse my American accent.' However, he can change the programme and alter the accent. On special occasions he likes to use one with a Scottish burr, which is perhaps the closest he can get to his natural voice. Timothy Hawking thinks his father's new voice suits him. Of all the family he is the one who can least remember Stephen's real voice, as he was only six at the time of the operation in Switzerland, and there had been very little voice left for many years before then.

With his new voice and a degree of financial security, a few weeks after leaving hospital Hawking was able to resume work on the manuscript. In collaboration with Peter Guzzardi, and taking on board suggestions quietly solicited from other readers, they decided to scrap a number of sections and rewrite some others. Hawking wanted to add a mathematical appendix, which would list the equations forbidden in the text, but Guzzardi vetoed the idea. 'It would terrify people!' he said.

As the two men worked on the manuscript and the publicity machine at Bantam began to get into gear for publication in the spring of 1988, Al Zuckerman was not idle. Having sold the rights for America and Canada, he was keen to find buyers for the rest of the world. Publishers in Germany and Italy both offered advances of $30,000 without even seeing the manuscript, and there was growing interest from Japan, Scandinavia, France and Spain. To his surprise he even received offers from Korea, China and Turkey, and two from Russia – a country to which he had never before managed to sell a book. 'I had two offers from publishers in Moscow,' he told the *Bookseller*. 'They don't compete – they both made the same offer.'[11] It seemed everyone wanted to get in on the action.

Global interest in Stephen Hawking's book was exceeding

Zuckerman's most optimistic dreams. Only in one major country did he encounter a problem: Britain.

British publishers were the most sceptical I encountered. When I showed the earlier version in the UK, Dent offered £15,000 and there were other offers of £5,000 and £10,000. I didn't think they were serious enough, so I withdrew.[12]

For the meantime, there was no UK publisher for a book by a British author that had been taken in almost every other country in the world.

No further progress was made until the American Booksellers Association convention in 1987. Mark Barty-King of Bantam UK had heard of the book through company connections. Bumping into Zuckerman at the convention, he asked him if he could read the manuscript. After reading it, he arranged to meet Zuckerman to declare an interest in the book. Zuckerman told him that he wanted £75,000 for the UK rights. Barty-King suddenly lost his enthusiasm:

£75,000 [at that time] seemed an outrageous advance for what was a *difficult* book, although a very distinguished one. Whether he tried it out on other people I don't know, but eventually we decided we could come up with £30,000. He said he had to try other publishers; we said, 'If you do, we want the floor.'[13]

Penguin, Collins, Century Hutchinson and others all failed to meet the floor of £30,000. Zuckerman returned to Bantam UK and accepted their offer.

Even then, Mark Barty-King's final decision was touch and go. The evening before he presented the idea to the scheduled editorial meeting, he sat down to do some sums. As usual he began to calculate projected sales. Hardback: home 3,000 copies, stock 2,000, export 500; trade paperback: 10,000 copies, stock 10,000, export 3,000; Australia and New Zealand: 3,000. The calculations didn't add up. Finally he added £5,000 for serial rights within the UK and he could just about justify the acquisition. He took it to the meeting and, against all advice from his colleagues, forced it through. The company would not make a penny from this book, he was convinced of that. However, on the plus side, a prestigious book such as this could only enhance their profile as publishers of 'serious' books and if they did not actually lose money from the deal, then it would be worth taking the risk.

It was not until he met Hawking in person at the Frankfurt Book Fair the autumn before publication that Mark Barty-King began to get an inkling of the man's enormous presence:

It's only when you meet him that you realize how extraordinary he is. What in particular comes as such a surprise, after all he has been through, is that you get such a strong impression of his sense of humour.[14]

After signing up the book, he told a journalist:

It's a book by one of the greatest minds of our time, discussing the elemental subject of who we are and where we come from. It is a lucid and very personal book, one which I personally found quite difficult to read because of the subject-matter, but one which I considered to have enormous appeal.[15]

In Frankfurt, Hawking delivered a short talk to the gathered publishers in a library hired for the occasion. He described his life as well as the philosophy and motivations behind the book. According to Guzzardi, they were enthralled. In the lead-up to publication in the USA, Guzzardi had a series of meetings with Bantam's Director of Marketing to discuss exactly how they would approach the promotion of the book.

Years earlier, when Simon Mitton learnt that Hawking had signed to a major trade publisher, he had given Hawking a piece of friendly advice. 'Do be careful if you're dealing with those people, Stephen,' he had said. 'Do ensure that you are quite certain that, if the aim is to make money and sell lots and lots of books, you don't mind the marketing techniques.'

'What do you mean?' Hawking had asked.

'Well, I wouldn't put it past them to market it as "Aren't cripples marvellous?". You've got to go into it with your eyes open. If you don't mind that approach, OK.'

In the event, Mitton's advice was unfounded. Guzzardi had no intention of promoting the book in the way Mitton had feared:

> We could have gone two ways. We could have Bantamized it – planes over Manhattan with sky-writing, T-shirts, etc., or we could go classy, tasteful, the quality-rap. The author is prestigious, we thought. Put marketing muscle into it, but do it tastefully. That was the alternative, and that's what we decided to do.

Less than a month before publication, Hawking received a surprising phone call from his agent Al Zuckerman. Peter Guzzardi, who had seen the book through from the beginning, had told him that he had been offered his own imprint at Crown and was leaving Bantam. The final stages of carrying the book through promotion and the nervy early-sales period would be handed over to a new editor. One of the last decisions Guzzardi made about the book was the final choice of title. Hawking thought that *A Brief History of Time* might come across as a little too frivolous, and had misgivings about the word 'brief'. It was Peter Guzzardi who managed to convince him that it was a brilliant title, succinct but definitive. According to Guzzardi, what finally convinced Hawking was when he remarked that the word 'brief' in the title made him smile. 'Stephen saw the point immediately,' says Guzzardi, 'he likes to make people smile.'

When handed the portfolio for this strange, difficult book, *A Brief History of Time*, the new editor at Bantam got cold feet. The new editor's first decision was to reduce the book's first print run drastically – to 40,000.

*

A Brief History of Time: From the Big Bang to Black Holes hit stores all over America in the early spring of 1988. The launch party took place at the Rockefeller Institute in New York, where a banquet was held in the author's honour and Hawking gave a short speech to promote the book. According to the other guests, after a long day of celebrations and seemingly endless introductions and meetings, Hawking was still full of energy and in a party mood.

The gathering moved to the embankment overlooking the East River. Stephen was in fine form. The years of work on *A Brief History of Time* were finally over, and the book was in the shops and would, it was hoped, do well. Friends remember how he wheeled around from guest to guest in very high spirits. There was a definite buzz of excitement in the air. It was a clear night, the stars shone brightly over the river and the city lights were reflected in a spectrum of coloured points in the water. Glasses were continually refilled, and although Hawking himself can drink very little alcohol and has no real sense of taste, by all accounts he seemed intoxicated by the atmosphere. There were some anxious moments, however. One friend recalled that both Jane and Stephen's nurse were terrified that in his excitement he would roll his wheelchair into the river.

Late in the evening a small group of close friends and family returned to the hotel. As they passed through the lobby, Stephen noticed a dance going on in a ballroom nearby. Insisting that it was too early to go to bed, he wheeled himself off in the direction of the music, intent on crashing the party. His friends were dragged along and persuaded to join in, and Hawking ended the celebrations in the early hours, whirring around the dance floor with the band playing on long after the original party had ended.

Bantam carried through their plan of a low-key launch for the book. There were no pre-arranged window displays or huge posters of the author. Pre-launch indicators from sales reps were encouraging but confused. Shops were keen to take the book, but did not know quite where to put it or what to do with it. Then, days after publication, near-disaster struck. An editor at Bantam, looking through a copy from the first print run, noticed that two of the pictures were in the wrong place. There was instant panic. The 40,000 print run was already in the shops. Sales staff hurriedly began to phone the larger bookshops.

'We've made a mistake,' they said. 'We'll have to recall all your copies.'

To their amazement, there were no unsold copies left. Shops all over America had already filled in re-order forms for more.

According to executives at Bantam, this was the first sign that they were on to something really big. Wasting no time, a reprint of the corrected version was ordered immediately and rushed to retailers as quickly as possible. Much to Bantam's delight, *Time* magazine ran a large article about Hawking in the month of publication, and favourable reviews began to appear in quality newspapers and magazines across the States. Within weeks of publication *A Brief History of Time* entered the best-seller list and climbed effortlessly to the top.

Suddenly, window displays appeared in bookshops all along Fifth Avenue, and posters of Stephen Hawking were put up over shelves packed with his book in shops all over America.

The cover of the American edition of the book shows Hawking sitting in his wheelchair against a backdrop of stars. He looks very stern and is staring at the camera, almost frowning. Hawking has said that he was always unhappy with this picture but that he had no say in its use. Some of his friends and family thought that the picture did not really express his true character and lacked humour.

One book reviewer took exception to Bantam's putting a photograph of the author in a wheelchair on the front cover, declaring it to be exploitative, a cynical commercial move on the part of the publisher to get the most mileage possible from their crippled author.

Peter Guzzardi was deeply offended by the suggestion. 'It was obvious the reviewer didn't know Stephen, to think that he could be exploited,' he said. 'No one could exploit Stephen Hawking. He is quite capable of looking after himself.'

'I think the reasoning behind that guy's comments was pathetic,' Guzzardi recalled with disgust on another occasion. 'It was a triumph for a man in Hawking's physical condition to be on the cover of his own book. It's inspiring.'

By the summer of 1988, Stephen Hawking's 'difficult' book had stayed in the best-seller list for four months and had sold over a half a million copies in America. He was very rapidly becoming a household name. The publishing phenomenon of the year hit the national news – and every airport bookstall in the country.

In Chicago, a Stephen Hawking fan club was hurriedly organized and started selling Hawking T-shirts. Among the 'science set', he began to achieve the status and commercial trappings of a rock star in schools and colleges from LA to Pittsburgh. The schoolboy who had been a devoted fan of Bertrand Russell was now, some thirty years later, himself a schoolboys' hero.

June 1988 saw the British publication of *A Brief History of Time*. It immediately followed the same pattern of instant success it had enjoyed in America. Bookshops sold out every copy within days. A few days after publication, one of us (M.W.) searched every bookshop in Oxford and London and could not find a single copy left on sale. Weeks later, he tracked down a copy – the last remaining in the bookshop at the World Trade Center in New York.

British sales reps were reporting overwhelming interest from retailers the length of the country. Waterstones in Edinburgh wrote to the publisher to say that they wanted to mount a window display and were planning to order 100 copies of the book. But despite the obvious interest the book was generating, the British publisher was slow to appreciate the scale of its success. Mark Barty-King at Bantam UK had decided to increase the first print run from 5,000 to 8,000, but these were sold by the end of their first day in the shops. Once again an immediate reprint was ordered. By the beginning of 1991, *A Brief History of Time* had gone to twenty reprints in Britain and was still selling an average of 5,000 copies a month in hardback. The book was leaving the book-shelves faster than the printers could produce new copies. The buyer for W.H. Smith was quoted as saying, 'Demand for the book had completely outstripped what we were expecting. It has almost become a cult book.'[16]

Reviews appeared in publications ranging from *Nature* to the *Daily Mail*, all of them favourable. Interview after interview appeared in newspapers and magazines. Hawking was becoming such a celebrity that he had to pick which journalists he would talk to.

'It was interesting to see the interviews he went for,' said Wendy Tury at Transworld. 'He wanted to do the *Sunday Mirror*, for instance.'[17]

Hawking's attitude was that he wanted to reach the broadest audience possible. He wanted plumbers and butchers to read his book as well as doctors, lawyers and science students:

> I am pleased a book on science competes with the memoirs of pop stars. Maybe there is still hope for the human race. I am very pleased for it to reach the general public, not just academics. It is important that we all have some idea of what science is about because it plays such a big role in modern society.[18]

Entering the *Sunday Times* best-seller list within two weeks of publication, it rapidly reached number one, where it remained unchallenged throughout the summer. At the time of writing, it is still in the top ten of the UK hardback best-seller list, and has still not been issued in paperback in Britain. The only book to have enjoyed greater longevity in the list is *The Country Diary of an Edwardian Lady*. By 1990, it had beaten off all other rivals, including Jacob Bronowski's phenomenally successful *The Ascent of Man*, which enjoyed over two years in the chart in the early seventies.

People began to stop Hawking in the street and proclaim their deepest admiration. Timothy was said to be embarrassed by such incidents, but Stephen revelled in it. One reviewer compared *A Brief History of Time* to *Zen and the Art of Motorcycle Maintenance*.* Family and friends were horrified, but Hawking took it as a compliment – a clear sign that he had succeeded in reaching his target audience.

Reviewers and commentators seemed bemused by the book's success. John Maddox, the editor of *Nature*, wrote towards the end of 1988:

Those who worry about the supposed public ignorance of science must surely be comforted to know that in the United States there are now in circulation 600,000 copies of Professor Stephen Hawking's book *A Brief History of Time* . . .

Curiously, among roughly a score of people I have questioned during a visit to California (not all of them scientists), I found none who did not know of the book, three who owned a copy and none who had yet read it. This seems odd for a volume of only 198 pages whose author's estimate of the reading-time can be inferred from his statement that 1,000 calories of nutriment will be required to capture its information content, roughly half a day.

Indeed, there is a strange embarrassment about the book. People say it is a 'cult' book, or describe Professor Hawking as a cult figure. In California, well-used to the coming and going of gurus differing in persuasions and persuasiveness, this explanation may seem natural. But even California cannot have absorbed all 600,000 copies.[19]

* A cult success of the seventies.

In August 1988, Simon Jenkins of the *Sunday Times* wrote:

I am all but mystified. Throughout this summer, a book by a 46-year-old Cambridge maths professor on the problem of equating relativity theory with quantum mechanics has been on the British non-fiction best-seller list. For the past month it has been top. Michael Jackson and Pablo Picasso have been toppled. Stephen Hawking's *A Brief History of Time* has notched up five reprints and 50,000 copies in hardback. This is blockbuster territory.[20]

Everyone, including many of the people who put it in the best-seller list, seemed startled by the book's cosmopolitan appeal. It was obvious that Hawking had indeed managed to achieve the accolade of having plumbers and butchers buying his book. There were simply not enough science students in the world to account for the sales figures. One writer recounted a story about a scientist who stopped at a garage in America and began to chat to the service attendant. When the attendant discovered the driver was a scientist, he asked, 'Do you know Professor Hawking? He's my hero.'[21] Suddenly everyone was a Hawking fan, and everyone had a pet theory as to how the book had become such a remarkable success.

So, what is the secret of its success? It is a question still being asked years after *A Brief History of Time* took up residence in the best-seller list.

In April 1991, nearly three years after its British publication, a tiny article appeared in the gossipy 'Weasel' section of the *Independent* magazine which questioned how many people had actually read the book:

That brilliant man Mr Bernard Levin has admitted in his *Times* column that he is unable to get beyond page 29 of *A Brief History of Time* by Professor Stephen Hawking. This raises a question: if the brilliant Mr Levin can only get as far as page 29, how is the average punter likely to fare as he embarks on the quest for knowledge about the origins of the universe?

Yet the fact remains that this slim scientific treatise priced by

Bantam at £14.99 has sold 500,000 copies in this country alone, and that come July it will have been in the best-sellers list for three whole years. Understandably, the publishers have no plans to issue a paperback edition.

How does one explain the extraordinary success of a book which so few of its purchasers are able to understand? Amateur psychiatrists point to the author's condition. He is a victim of motor neuron disease who was given up by his doctors years ago. Yet, against all the odds, he wrote his book. It is a heroic tale, but is it enough to explain the book's success?

I do not think so. Nor will it do to say that readers hope to discover the truth about the origins of the world in which they live. The word will have got round by now that there is no easy answer. The mystery of the book's success is by now almost as baffling and fascinating as the mystery of the origins of the universe. I am prepared to offer a small prize (say £14.99) to any reader who can provide an explanation that is at all convincing.[22]

The article provoked a flood of letters, including one from Hawking's mother, Isobel, published the following week, in which she wrote:

Sir: I have to declare an interest, as I am the mother of Professor Stephen Hawking, but I have given some thought to the reasons for the success of *A Brief History of Time* . . . a success which surprised Stephen himself. I believe the reasons to be complex, but shall attempt to simplify them – as I see them.

The book is well-written, which makes it pleasurable to read. The ideas are difficult, not the language. It is totally non-pompous; at no time does he talk down to his readers. He believes that his ideas are accessible to any interested person. It is controversial; plenty of people oppose his conclusions on one level or another, but it stirs thought.

Certainly his fight against illness has contributed to the book's popularity, but Stephen had come a long way before the book was even thought of. He did not collect his academic and other distinctions because of motor neuron disease.

I do not claim to understand the book myself, though I did read it to the end before coming to this conclusion. I think my age and type of mental training have something to do with my non-comprehension. Without wishing to doubt the brilliance of Mr Levin's intellect, I should hesitate to assume from his non-comprehension that most people share it.[23]

Isobel Hawking's last point seems to have got to the root of the matter perfectly. While some would consider the classical 'Oxbridge arts' education the perfect foundation for later identification as 'an intellectual', there are other forms of education which, as we rush headlong towards the twenty-first century, may be more appropriate for the 'intellectuals' of the future. Ask any scientist about the prejudices of the scientifically untrained. Such people make themselves known at any normal dinner party. The sociable scientist has a surfeit of sorry tales of how the uninitiated protect their own ignorance with Levinesque belittlement, almost revelling in the fact that they don't understand scientific matters. It is often easier to make a joke of things you do not wish to admit to than to be honest and confront them. In Britain, especially, this xenophobia is nurtured by the vestiges of Victorian images of the scientist as little more than a workman dirtying his hands in a laboratory, messing around with chemicals and bizarre-looking instruments.

Among the other replies to the 'Weasel' piece was another letter which neatly exposed such misplaced intellectual snobbery:

Sir: You are mistaken in thinking that few of the purchasers of *A Brief History of Time* are able to understand the work. It is only those who, like Bernard Levin, have had a limited education who have this problem.

My 17-year-old son, a physics A Level student, found the book very easy to understand and wished that Stephen Hawking had written in greater depth. This is a boy who never reads a novel and usually buys only the *Sun*. He would himself claim the £14.99

offered by the Weasel to those who could explain the popularity of Hawking's book, but he is hardly capable of constructing a letter.

This bears out the theory . . . that there are different sorts of intelligence. Just as there are philistine scientists there is an arts intelligentsia which is mathematically and scientifically illiterate. Never mind Shakespeare: perhaps schools should be teaching concepts that help one to understand the very basis of the nature of the Universe.[24]

Despite such forthright opinions, a great many people believe that *A Brief History of Time* has turned out to be *the* book to be seen with in the eighties and nineties. Soon after publication several articles appeared in which the writer commented on the fact that friends and colleagues were in competition to see how far they had managed to get through it. Both the writers of this book have compared notes on the comments of our non-scientist friends (and sometimes scientifically trained ones too) who claim over dinner that they are trying it 'a page a day' or that they are 'three pages further on than my next-door neighbour'. Even Simon Jenkins, who displays a continuing high regard for Hawking and his book, waded in with:

Hawking is, I am sure, benefiting from 'wisdom by association'. Buying a book is a step more virtuous than merely reading a review of it, but need not involve reading it. On the coffee-table or by the loo, a book is the intellectual equivalent of a spare Gucci label stitched on a handbag or an alligator on a T-shirt.[25]

Others have claimed that *A Brief History of Time* has sold so well because it has been latched on to by a lost generation of post-yuppie Greens who see it as a symbol of new-age wisdom, that it somehow takes on semi-religious importance in their minds. Of course, Hawking finds such notions hilarious.

So, what do Hawking's colleagues think of his book? If the truth be known, many have not read it, claiming that they hardly see it as a beach-read. Among those who have, there is a variety of opinions. A number have drawn the conclusion that Hawking did not go far enough and that the book should have been twice the length, but that perhaps is the professional in them talking.

Some like it, others do not. More than one physicist has said that he felt Hawking was wrong to integrate accepted and established scientific conclusions with his own controversial speculations without informing the lay-reader of any distinction between the two. Others believe that Hawking's insistence on including potted biographies of Galileo, Newton and Einstein at the end of the book is pretentious – that it implies that the author thought the name 'Hawking' would be the next in line in any future *A Brief History of Time*. This last view seems at odds with the man's own opinion of the media-hype surrounding his status. For, he would claim, it is they, not he, who have made such proclamations. Others would argue that he has every right to think of himself in the same light as this illustrious triumvirate.

Whatever the reason for the book's amazing success, it has far outstripped the wildest expectations of the publishers who signed it up, the agent who saw its commercial value and, most of all, the writer and editor who created it.

The final story of its cosmopolitan appeal must be reserved for a tale from the Russian physicist, Andrei Linde. Soon after the book's publication, he was flying across America for a conference and happened, not unusually, to be seated next to a businessman. Some way into the flight he glanced across and noticed that the man was reading Hawking's book. Without having been introduced and before the usual small-talk, they struck up a conversation about it.

'What do you think of it?' Linde asked.

'Fascinating,' said the businessman. 'I can't put it down.'

'Oh, that's interesting,' the scientist replied. 'I found it quite heavy going in places and didn't fully understand some parts.'

At which point the businessman closed the book on his lap, leant across with a compassionate smile and said, 'Let me explain . . .'

15

The End of Physics?

Stephen Hawking is fond of suggesting that the end may be in sight for theoretical physics. Hearing Hawking tell you that physics may be coming to an end became something of a cliché in the trade in the 1980s, as at the beginning of that decade he used his inaugural lecture as Lucasian Professor to pose that question. Ten years on, the end doesn't look any closer than it did then, but he is still optimistically proclaiming it. But even if theoretical physics really did reach the 'end' Hawking so eagerly predicts, there would still be plenty of work left for physicists to do.

In an interview in *Newsweek* in 1988, Hawking said that after discovering a theory of everything 'there would still be lots to do', but physics would then be 'like mountaineering after Everest'.[1] Other cosmologists, including Martin Rees, prefer a slightly different analogy. They point out that learning the rules of chess is only the first step on a long and fascinating path to becoming a grandmaster. The long-sought-after theory of everything, they say, would be no more than the physics equivalent of the rules of chess, with grandmaster status still far away over the horizon.

The immediate goal of physics – the Holy Grail which Hawking and a few other researchers believe lies just around the corner – is a complete, consistent, unified theory in which all physical interactions are described by one set of equations. To see what this means, and how daunting the search for such a theory must be, we shall look at the modern understanding of the way the Universe works, which requires *four* separate theories to explain different features of the world.

Back in the nineteenth century, only two theories were needed (so in a way physics has got more complicated in the past hundred years). Newton's theory of gravity described the force which holds planets in their orbits around the Sun, or makes an apple fall from a tree; Maxwell's equations of electromagnetism described the behaviour of radiation, including light, and the forces that operate between electrically charged particles, or between magnets.

As we explained in Chapter 2, though, these two theories were incompatible. Maxwell's equations set a speed for light which is the same for all observers, while Newtonian mechanics said that the speed measured for light would depend on the motion of the observer. This dichotomy was one of the principal reasons why Einstein developed first the special theory of relativity and then the general theory – an improved theory of gravity that *is* compatible with Maxwell's equations. Both the general theory and Maxwell's theory are, however, 'classical' theories in the strict sense of the term. That is, they treat the Universe as a continuum. Space, in the classical view, can be subdivided and measured in units as small as you wish, while electromagnetic energy can come in a quantity as small as you wish.

The quantum revolution changed the way physicists view the world. They now regard the Universe as discontinuous, with an ultimate limit on how small a 'piece' of electromagnetic energy can be, and even on how small a unit of time or a measure of distance can be. It was discoveries concerning the nature of light that led to the quantum revolution, and electromagnetism was eventually superseded by a new theory, quantum electrodynamics (QED), that incorporates the best of Maxwell's theory with the new quantum rules.

But QED did not become established until the 1940s, by which time two 'new' forces were on the agenda. Both these forces have only very short range, and operate only within the nucleus of an atom (which is why they were unknown in the nineteenth century, before the nucleus was discovered). One is called the strong force, and acts as the glue which holds the particles in the nucleus together; the other is known as the weak force (because, logically enough, it is weaker than the strong force), and it is responsible for radioactive decay.

In many ways, however, the weak force resembles the electromagnetic force. Building from the success of QED, in the 1950s and 1960s physicists developed a mathematical theory that could describe both the weak force and electromagnetism with one set of equations. It was called the 'electroweak' theory, and it made one key prediction: with the weak force there should be associated three types of particle which, between them, play much the same role that the photon (the particle of light) does in QED. Unlike the photon, however, these particles (known as W^+, W^- and Z^0) should, according to the new theory, have mass. Not just any old mass, either, but very well-determined masses – about nine times the mass of a proton for the two W particles, and eight times the mass of the proton for the Z^0. In 1983, the particle accelerator team at CERN in Geneva found traces of particles with exactly the right properties. The electroweak theory was a proven success, and physicists were back to just three theories needed to explain the workings of the Universe.

With this success under their belts, theorists have developed a theory similar to QED to describe the strong force. We now know that nuclear particles (protons and neutrons) are actually made of fundamental entities known as quarks. Quarks come in different varieties, and physicists whimsically give these the names of colours – red, green, and blue. This doesn't mean that quarks really are red green or blue, any more than the fact that a drink is called a rusty nail means that it really does contain oxidized iron. They are just names. But, extending the whimsy, physicists call the quantum theory that describes how quarks interact, and which is responsible for the strong force, 'quantum chromodynamics' (from the Greek word for colour), or QCD. There are several promising ways now being investigated that might lead to a single theory that encompasses both QCD and the electroweak theory. Such sets of equations are known, rather pretentiously, as Grand Unified Theories, or GUTs. But QCD is not yet as well established as the electroweak theory, and the GUTs themselves are only indicative of the form a future, definitive theory might take.

Even worse, the pretentiousness of calling these 'Grand Unified Theories' is highlighted by the fact that none of this progress towards unification takes any account of gravity at all! The first force of nature to be investigated, and at least partially understood, it has proved the most intractable when it comes to trying to fit it into the quantum mould. Without gravity included in their mesh, it seems fair to say that – paraphrasing Hawking's famous comment about black holes – Grand Unified Theories ain't so grand, after all. In spite of Hawking's success in using a partial unification of quantum theory and general relativity in his investigations of black holes and the beginning of time, gravity is still best described by the general theory of relativity – a classical, continuum theory.

The prospect of incorporating gravity into what, we suppose, would have to be called a 'super-unified theory' has been 'just around the corner' for well over a decade. Logically, we might guess that first we need to develop a quantum theory of gravity, and then build from this to a unification with the other three forces. One feature of any such quantum theory of gravity is that it, too, must incorporate particles that are associated with the gravitational force, again reminiscent of the way photons are associated with electromagnetism. (In case you are wondering, yes, there are similar particles involved in QCD, the theory of the strong force; they are called 'gluons', but nobody has yet detected one.) Physicists even have a name for these hypothetical particles of gravity – 'gravitons'. But just as calling a quark 'red' does not mean that it is actually coloured red, so giving the quantum gravity particle a name does not mean that anybody has yet found one, or even that anybody has come up with a satisfactory quantum theory of gravity.

At the time of Hawking's inaugural lecture in 1980, researchers were getting excited about a family of possible quantum gravity theories that together go by the name of supergravity. One version of supergravity is called '$\mathcal{N} = 8$', because as well as predicting the existence of one type of graviton it also requires an additional eight varieties of particle known as gravitinos (together with a further 154 varieties of other as yet undiscovered particles). The plethora of particles associated with this favoured version of supergravity may seem unwieldy, and it is, but it does represent a considerable advance on previous attempts to find a quantum theory of gravity, which seemed to require an infinite number of 'new' particles. Indeed, out of all the variations on the supergravity theme, $\mathcal{N} = 8$ is the only one that operates naturally in four dimensions (three of space plus one of time) and contains a finite number of particles. It certainly got Hawking's vote as the theory most likely to succeed in 1980.

In the next few years, everything changed. By the mid-1980s, enthusiasm for supergravity had been swept away in a rising tide of support for a completely different kind of idea, known as string theory. The central idea of string theory is that entities that we are used to thinking of as points (such as electrons and quarks) are actually linear – tiny 'strings'. Such strings would be very small indeed: it would take 10^{20} of them, laid end to end, to stretch across the diameter of a proton. Such strings might be open, with their ends waving free, or closed into little loops. Either way, some theorists believe, the way they vibrate and interact with one another could explain many features of the physical world.

String theory actually dates back to the late 1960s, when it was invoked to describe the strong force. The success of QCD left this early version of string theory by the wayside, although a few mathematicians dabbled with it out of interest in the equations, rather than in any expectation of making a major breakthrough in unifying our understanding of the forces of nature. In the mid-1970s two of those researchers, Joël Scherk in Paris and John Schwarz at Caltech, actually found a way to describe gravity using string theory. But the response of their colleagues was, essentially, 'Who needs it?' At that time, most gravity researchers were more interested in supergravity. String theory wasn't needed to explain the strong force, supergravity looked promising, so why bother with strings at all?

Their attitude changed when it turned out to be horrendously difficult to do any calculations at all using the $\mathcal{N} = 8$ supergravity theory. Even if there were no infinities to worry about, 154 types of particle, in addition to the graviton and eight gravitinos, were almost too much of a handful to keep mathematical tabs on. Hawking says that it was generally reckoned in the early 1980s that, even using a computer, it would take four years to complete a calculation, checking that all the particles in the theory were accounted for, with no infinities still hidden away somewhere, and that it would be almost impossible to carry out the calculation without making a mistake. So nobody was prepared to give up their careers to do the calculation.

The main reason for the revival of interest in string theory in the mid-1980s, however, was the realization that in their most satisfactory form these theories *automatically* include the graviton. In other attempts to build a quantum theory of gravity, researchers had started out knowing the properties a graviton ought to have, and tried to build a theory around it, even if that meant taking 162 other particles on board as well. With string theory, they were working with the quantum equations in a general way, playing mathematical games, and found that the closed loops of string described by some of the equations have just the properties required to provide a description of gravity – they are, indeed, gravitons. Inevitably, the new variation on the string theme was dubbed 'superstring theory'. By 1988, with the publication of *A Brief History of Time*, it was *this* road towards super-unification that Hawking was enthusiastically endorsing.

But there are still snags. One is that people are still unsure what all the equations mean. As the example of the graviton illustrates, the equations have come first, with physical insight into their significance lagging behind, and there are still plenty of equations for which, as yet, there is no physical insight. This is quite different from the way the great developments in physics were made earlier in the twentieth century, and, indeed, in the centuries back to Newton's time. For example, Einstein used to tell how he was sitting in his office in Berne one day when he was suddenly struck by the thought that a man falling from a roof would not feel the force of gravity while he was falling. That insight into the nature of gravity led directly to the general theory of relativity – physical insight first, and then the equations. Exactly the same process was at work when Newton watched the apple fall from a tree, and went on to develop his theory of gravity.

But it seems that science, or at least physics, no longer works like that. One of the pioneers of superstring theory is Michael Green, of Queen Mary College, in London. In an article in *Scientific American* in 1986, he pointed out that with string theory details have come first; we are still groping for a unifying insight into the logic of the theory. For example, the occurrence of the massless graviton . . . appears accidental and somewhat mysterious; one would like them to emerge naturally in a theory after the unifying principles are well established.[2]

Another oddity of superstring theory does not seem to trouble the mathematicians, but demonstrates all too clearly to lesser mortals how far these ideas have strayed from everyday reality. What appeared to be the best versions of superstring theories, the ones in which gravitons seem to emerge naturally (if mysteriously) from the equations, only work in a little matter of twenty-six dimensions. So, if superstrings really do describe the workings of the Universe, where are all the extra dimensions hidden?

Mathematicians, in fact, have little difficulty in disposing of 'extra' dimensions of space. They use a trick which they call 'compactification', which can be understood by looking at the appearance of objects viewed from different distances in the everyday world. The standard image which they ask us to conjure up is that of a hosepipe. Viewed from close up, it is clear that a hosepipe consists of a two-dimensional sheet of material wrapped around a third dimension. But if we move back from the pipe and study it from far away, it looks like a one-dimensional line. If we look at this one-dimensional line end-on, it even looks like a point, with zero dimensions.

Taking a slightly different example, we all know from everyday experience that the surface of the Earth is far from smooth – it has wrinkles and bumps that we call valleys and mountains, so extreme that in some places it is impossible to walk across the surface. Yet to an astronaut far out in space, the surface seems to be very smooth and regular.

This may be why we do not perceive the other twenty-two dimensions of space. They may be curled up, or 'compactified' into the multi-dimensional equivalent of cylinders and spheres.

Each point of space that we perceive must really be a 22-dimensional knot of space, curled up very tightly so that we cannot see the bumps. How tightly? Roughly speaking, the complex structure of space would only be apparent on a scale of less than 10^{-30} of a centimetre. (For comparison, a typical atomic nucleus is about 10^{-13}cm across. So a nucleus is about a hundred million billion times bigger than the knots in the structure of space. In relation to a nucleus, the knots are one hundred thousand times *smaller* than a nucleus is compared with your thumb.)

Although mathematicians have no trouble describing such phenomenal compactification, it does raise the interesting question of why twenty-two dimensions should have rolled up in this way, while the other three dimensions of space have been expanding ever since the Big Bang. Intriguingly, both the familiar law of gravity and the equations of electromagnetism discovered by Maxwell only 'work' in a universe where there are three dimensions of space plus one of time. If, for example, there were more spatial dimensions, there would be no stable orbits for planets to follow around a central star. The slightest disturbance, and the planet would either fall into the star and be burnt, or drift away into space and freeze. In fact, as Hawking points out, there wouldn't even be any stable stars – any collection of gas and dust would either break apart or collapse immediately into a black hole.

So the laws of physics may be telling us that, *whatever* number of dimensions you start out with, all but three spatial dimensions and one time dimension must be unstable, and will compactify. There is even a hint, from some new research, that the collapse of the other twenty-two dimensions might have provided the driving force which started the other three dimensions expanding. And all of this, of course, relates to the idea of anthropic cosmology, which we described in Chapter 13. Perhaps there are other universes, other bubbles in spacetime, where the compactification worked out slightly differently, leaving, maybe, six or seven spatial dimensions (or only one). But since those universes will contain no suitable home for life, there will be nobody in them trying to puzzle out the nature of physics. If life-forms like us can exist only in a universe with three spatial dimensions, it is no surprise to find that the Universe we live in does indeed have only three spatial dimensions!

So how close is the study of physics to answering the ultimate questions of life and the Universe? Will there be no work left for theoretical physicists to do in the twenty-first century?

In 1980, in his Lucasian lecture, Hawking suggested that we might see the end of physics 'by the end of the century'. By this, he meant that physicists would have a complete, consistent and unified theory of the physical interactions that describe all observable phenomena. Something along the lines of superstring theory, perhaps.

As Hawking acknowledged, there have been previous occasions on which physicists have thought they were on the brink of finding all the answers. Most famously, at the end of the nineteenth century there was a general feeling that, with Maxwell's and Newton's equations firmly established, everything else would be merely a matter of detail, a question of dotting the i's and crossing the t's of science. Hardly was this feeling firmly established, than physics was turned on its head by the twin revolutions of quantum theory and relativity theory. And yet, by the late 1920s – just a generation later – the pioneering quantum physicist Max Born was telling people that there would be nothing significant left for theoretical physicists to do within six months.

At that time, the only fundamental particles known were the electron and the proton, and it seemed to Born that they were well understood. In the early 1930s, however, the neutron was discovered, and we now know that both the neutron and proton are made of yet more basic particles, the quarks.

Even taking Hawking's optimism of 1980 at face value, though, this would not mean that all physicists would be unemployed after the year 2000. As Hawking emphasized in that lecture, the laws of physics that Born was so proud of more than sixty years ago really are all that we need, in principle, to describe the behaviour of chemical reactions. Biological processes, in turn, depend on the chemistry of complex molecules. Chemistry depends almost entirely on the properties of electrons, and in the 1920s Paul Dirac found a quantum equation that exactly describes how electrons behave. The snag is, this equation is so fiendishly complex that nobody has been able to solve it, except for the simplest possible atom (hydrogen), which has a single electron orbiting a single proton. In Hawking's words, from that Lucasian lecture:

> although in principle we know the equations that govern the whole of biology, we have not been able to reduce the study of human behaviour to a branch of applied mathematics.

Even if we had a genuine unified theory that contained all the forces of nature, it would be far more difficult to use this to work out the behaviour of the entire Universe than it is to work out your behaviour using Dirac's equation. So there is plenty of work left for theoretical physicists to do.

By the time *A Brief History of Time* appeared in 1988, Hawking was being more cautious about the end being in sight for theoretical physics. He talked of 'if' we discover a complete theory, not 'when'. Indeed, although the millennial resonance of the possibility of discovering a complete theory by the year 2000 obviously appealed in 1980, this is one of those prospects that keeps receding into the future. As we have said, physicists have been talking about such an end to physics being 'just around the corner' for at least twenty years, and usually, if pressed, they would say that the corner they expect to turn lies about twenty years ahead – *whenever* you ask them that question! As we enter the 1990s, even the most optimistic physicist now sets the date for finding a complete theory no earlier than about 2010, and most refuse to be drawn into such speculations.

Perhaps, though, they should regard the question of finding the ultimate theory with some urgency. For, at the end of his Lucasian lecture, Hawking made another forecast, and one which has stood the test of time (so far). Commenting on the rapid developments being made with computers during the 1970s, he said that 'it would seem quite possible that they will take over altogether in theoretical physics' in the near future. That hasn't quite happened yet. Although progress with computers has been even more dramatic in the 1980s than in the 1970s (for example, we are writing these words using computers more powerful than those available to a whole university full of mathematicians in the 1970s), computers still have to be directed in their efforts by human scientists. But complex problems such as calculations involving 26-dimensional strings would be inconceivable without the aid of computers. It is, perhaps, more likely that the computers will no longer need human direction in tackling these problems by the end of the present century, than that human physicists will have found their long-sought ultimate theory. The most prescient comment of all in Hawking's inaugural lecture may in fact have been his very last sentence, one which makes a suitable ending for our own discussion of his contribution to science:

Maybe the end is in sight for theoretical physicists, if not for theoretical physics.

16

Hollywood, Fame and Fortune

From conception to best-seller list, *A Brief History of Time* took over five years. During the same period, Hawking had continued his research and administration of the DAMTP. In 1984, long before the first draft had been completed, Hawking went on a lecture tour of China. The itinerary for the trip would have been strenuous for an able-bodied man, but he insisted on cramming in as much as possible during the visit. He motored along the Great Wall in his wheelchair, saw the sights of Peking, and gave talks to packed auditoria in several cities. Dennis Sciama has said that he believed the trip took a lot out of Hawking, and has even suggested that it helped precipitate his subsequent illness in Switzerland less than a year later.

However, there were other exertions along the way. In the early summer of 1985, Hawking undertook a lecture tour of the world. One of the most important stop-overs was at Fermilab, Chicago. At the core of the cosmology group at Fermilab were three larger-than-life characters, Mike Turner, David Schramm and Edward Kolb, who have perhaps contributed as much to legend and anecdote surrounding the global cosmology fraternity as they have hard science.

Mike Turner is a tall, handsome Californian with a voice indistinguishable from Harrison Ford's. His office at Fermilab, where he spends most of his working life, is filled with toys and gadgets. Hanging from the ceiling are inflatable airliners and UFOs. The walls are plastered with postcards from friends around the world, humorous messages and wacky pictures, the floor littered with books and boxes of scientific papers. One wall is taken up by a blackboard covered in the hieroglyphs of physics; another opens onto a view of the lakes and woods surrounding the massive concrete columns of the central building which splay at the bottom and converge at the top to form an inverted V.

Edward Kolb, known as 'Rocky' because of his penchant for fighting, is a cosmologist from Los Alamos who joined the cosmology group at the same time as Turner in the early eighties. He and Turner became great friends and gained a reputation as a comic duo at Fermilab, forever playing practical jokes and initiating mischief. Their lectures were invariably witty, entertaining occasions, Turner's often featuring brightly coloured cartoons of Darth Vader to illustrate his ideas.

The cosmology group was set up by David Schramm, who is Chairman of the Astronomy Department of the University of Chicago, a close friend of Hawking and a formidable personality on the international cosmology scene.

Hawking arrived at Fermilab to give a technical lecture to a large group of physicists from around the globe and promptly discovered that there was neither elevator nor ramp to enable him to reach the lecture theatre in the basement. Turner recalls how he and Kolb were escorting Hawking into the building when the horrifying thought suddenly struck them: how were they to get Stephen to the stage? They looked at each other and, without saying a word, Turner lifted Hawking's featherweight body into his arms and Kolb grabbed the wheelchair. Halfway down the aisle of the lecture theatre, Turner became aware that the entire audience was watching agog as they struggled to the stage, and suddenly remembered how Hawking hated to have attention drawn to his disabilities. In the event Stephen said nothing about the incident, realizing, he mentioned later, that there was absolutely no alternative.

Next day he gave a public lecture in Chicago, receiving a rock star's reception. The standing-room-only audience packed the auditorium and a number of people had to be turned away. He was recognized everywhere he went, and people stopped him in the street to express their interest in what he was doing. The title of his lecture was 'The Direction of Time'. To a startled audience he declared that, at some point in the distant future, the Universe would begin to contract back to a singularity and that during this collapse time would reverse – everything that had ever happened during the expansion phase would be re-enacted but backwards.

There were many who opposed Hawking's ideas, including his close friend Don Page. Indeed, Hawking himself knew that he was venturing into wild country. After the visit the two of them wrote opposing papers, published in the same issue of the scientific journal *Physical Review*. Hawking's paper led off the pair and concluded by saying that Page had some interesting arguments on the subject and that he may well be right. Eighteen months later, in December 1986, Hawking returned to Chicago to deliver a talk which announced that he had been wrong in 1985, and now proclaimed the opposing view to be correct: time would not go into reverse as the Universe contracted.

By this time Hawking and Guzzardi were tidying up the manuscript for *A Brief History of Time*, which Al Zuckerman was selling to foreign publishers, and Hawking himself had grown accustomed to his computer-generated voice synthesizer. A Cambridge computer engineer called David Mason had designed and built a portable version of the device operated by a mini-computer which could be attached to Hawking's wheelchair. Now his voice could go with him everywhere he went. He began to deliver lectures with the new machine in 1986. Suddenly, audiences could fully understand what he was saying, and although the voice did not produce sentences with the Home Counties accent Hawking would have preferred, what he had to say was so much clearer now that he no longer needed to use an interpreter.

Attending a Hawking lecture is, initially, a very odd experience. He is wheeled on to the stage by an assistant, his voice synthesizer is plugged into the public-address system, and the computer disks containing the text of his talk are inserted into the computer perched on the arm of his wheelchair. To the audience, Hawking looks totally passive, immobile but for facial expression, the tiny, imperceptible movements of his fingers operating the computer. He lifts his eyebrows and smiles at appropriate points; his eyes glint in the stage lights as his head lolls onto his chest. Standing in the wings are two nurses and a small group of research students, always ready to come to his assistance if needed. After an introduction by the organizer, and when the applause dies down, a disembodied voice suddenly bursts into the room from the PA speakers: 'In this lecture, I would like to discuss . . .'. The pre-programmed disks can hold a little under half an hour of his lecture, so that at a pre-designated point in the talk he has to announce to the audience that he is reloading his computer and will continue in a few moments.

After the talk he invites the audience to ask questions, but warns that the responses will take some time for him to programme into his computer. 'During this time,' he says, 'please talk among yourselves, read newspapers, relax.' The answers can take up to ten minutes to come back. A spokesman announces that Professor Hawking is now ready to reply, and the audience falls silent. There is no possibility of any interaction between the questioner and Hawking: the answer is accepted and the next person is already up and ready with another question. Sometimes Hawking's answer is a simple 'Yes' or 'No', a response which comes quickly.

Sometimes, just for fun, he has been known to deliberately wait five minutes before responding with a monosyllabic reply. The audience love it and burst into spontaneous laughter. On more than one occasion he has been known to wait five minutes, only to ask for the questioner to repeat the question. As he has grown older, Hawking's innate sense of mischief has not diminished in the slightest.

In December 1990 he was invited to deliver a public lecture at a symposium, held in Brighton. The venue was a huge complex of auditoria called the Brighton Conference Centre. Unfortunately for the delegates, the complex had to be shared with the rock group Status Quo who were performing in one of the main halls. Between five and seven o'clock, the intense concentration of audiences in various rooms and theatres around the building was broken by the band sound-checking in the Main Hall.

Interspersed with talk of worm-holes and neutron star astrophysics came the thump, thump, thump of a bass drum and the yells of roadies bellowing down microphones, 'One, two; one, two; testing, testing; one, two . . .'

On the evening before Hawking's talk, he was expected at an unofficial meeting in his hotel room at 8.30. At the appointed time, a small group of journalists and friends arrived, were let in, and sat down to wait for him. Twenty minutes later, Hawking's mother Isobel walked in, looking surprised to find them there.

'Where's Stephen?' one of the journalists asked. 'He was supposed to be here at 8.30.'

'Stephen? He's gone to see Status Quo,' Isobel replied.

A group of Hawking's students had wanted to see the band and had sent a representative to find out if there were any remaining tickets. Hearing that the concert had sold out months ago, the student had told the organizers that Stephen Hawking was next door and really wanted to see Status Quo. Within five minutes he was handed a few complimentary tickets. According to one of his students, Hawking thoroughly enjoyed himself and stayed throughout the whole concert.

After the publication of *A Brief History of Time*, there was a subtle shift of atmosphere at the DAMTP in Cambridge. There were incessant requests for interviews from newspapers and magazines from around the world. On several occasions over the next two years, a television crew took over the building to make a documentary about the life of the man who had become the most famous scientist in the world. The same stories appeared over and over again in a variety of languages, all telling of his courage in overcoming a crippling disease to become a scientific giant as well as a media hero. Journalist after journalist visited the cluttered office in Silver Street to spend an inspiring hour with the public's latest hero. Returning to their offices, they wrote about the drab paintwork at the DAMTP, the scruffy assistants, the ever-present nurses and the Marilyn Monroe poster pinned to the back of Hawking's office door.

Despite the countless thousands of words written about him, very little new information about the man appeared in the pages of the world's press. The details of ALS and the succession of awards and honours bestowed upon him were trotted out time and again, but Hawking was determined to maintain a degree of privacy amid the whirlpool of hype.

In the United States, ABC profiled Hawking in its *20/20* series, while in Britain a new documentary appeared called *Master of the Universe*, which won a Royal Television Society award in 1990. In the film, Hawking was shown bowling along the streets of Cambridge and in one shot was seen entering the main entrance of King's College. The autumn after the programme was televised, the admissions officer at King's was astonished to find a huge increase in the number of applications to study mathematics at the college. The television audience had obviously assumed that Professor Hawking taught and worked at King's College. In fact, he simply used a route through the grounds of King's as a convenient short cut for his wheelchair on the way to the DAMTP. But King's did not disabuse the bright young mathematicians suddenly eager for places at the college.

Hawking enjoyed the adulation and celebrity. He continued to travel around the world. The invitations to give public lectures were becoming overwhelming, and he could have spent his whole time delivering them unless he carefully selected the ones he would attend and those he could not. In Japan he was received as an idol, getting the sort of reception usually reserved for heads of state or internationally famous rock stars. Hundreds queued to hear him speak in lecture theatres throughout the country.

Back in Cambridge, the volume of mail Hawking received daily had long since become too much for him to handle personally. A research assistant and his secretary were given the responsibility of sifting through the piles of invitations, letters, documents and professional correspondence. For some years he had been receiving 'crank mail', a drawback of the job and experienced by many other famous scientists throughout the world, especially physicists. However, by the late eighties Hawking was beginning to receive an inordinate quantity of bizarre letters spanning the entire spectrum of eccentricity. Correspondents ranged from amateur physicists in country villages proposing ridiculous solutions to cosmological questions, to religious extremists criticizing what they saw as the intrusion of science into sacred areas. Before long, a 'cranks file' was set up at the DAMTP where the best examples of the genre were kept for entertainment value; the rest were put in the wastepaper bin.

Meanwhile, academic accolades and public acknowledgements of his scientific work kept coming. As early as 1985, long before the publication of *A Brief History of Time*, his portrait, commissioned by the trustees of the gallery, was hung in the National Portrait Gallery in London. In the late eighties alone he received five more honorary degrees and seven international awards. In 1988 he shared the Israeli Wolf Foundation Prize in physics with Roger Penrose for their work on black holes.

In January he travelled to Israel to receive the prize and a cash award of $100,000 at a ceremony at the Knesset, Israel's parliament in Jerusalem, attended by the Israeli president and other political and scientific figures from around the world. The award did not pass without controversy. Jewish legislators boycotted the event, claiming that Hawking's theories went against a tenet of Judaism that neither time nor objects existed before God created the Universe. Despite the protests, Hawking himself was pleased with the award, and in a typically double-edged comment, he told the press, 'I am very pleased. It shows that British science is still good, despite the government cuts.'[1]

In 1989 he was again honoured by the Queen when he was included in the Honours List for the second time. This time he was made a Companion of Honour, one of the nation's top honours, and attended a reception at Buckingham Palace the following summer to receive the award from the Queen. During the week when he officially became a Companion of Honour, he received a very rare accolade when Cambridge University made him an honorary Doctor of Science. Only in very special cases do academics receive honorary doctorates from their own universities. The award was presented by Prince Philip, Chancellor of the University, at a special ceremony in Cambridge. Hundreds of people lined the streets and applauded as Hawking wheeled along King's Parade in the procession of dignitaries, arriving at the Senate House to the accompaniment of the choirs of King's and St John's Colleges and the Cambridge University Brass Ensemble.

To complete an astonishing week, on the Saturday evening, as the sun set over the spires and towers of a Cambridge basking in the summer warmth, the strains of Bach, Vivaldi, Purcell and Handel could be heard as the Cambridge Camerata performed a special concert in Hawking's honour at the Senate House in the centre of the city. That night there was not a dry eye in the house, according to the local newspaper covering the event. As a special favour, the orchestra played Wagner's 'Ride of the Valkyries', one of Hawking's favourite pieces. As the applause for the musicians died down, Stephen wheeled up to the stage, turned and thanked the audience through his voice synthesizer, receiving a standing ovation from his friends and family, and members of the public there to honour the man who had achieved so much against all odds. According to one journalist:

> There were tears rolling down the cheeks of men and women as a tribute to his courage, as well as the exceptional brain which has continued to advance knowledge of time and space in spite of the ravages of a crippling disease.[2]

Another journalist told him at a reception after the concert that *A Brief History of Time* had received more inquiries from readers of the 'News' book page of his paper than any other book.

With Hawking's enhanced status as a world-famous scientist and writer, his campaigning for the rights of the disabled stepped up a gear. In 1989 a project was set up in Cambridge to create a special hostel for handicapped students at the University. It was called the Shaftesbury 'Bridget's' Appeal in memory of Bridget Spufford, the disabled daughter of a Cambridge history lecturer who had been unable to find a single university in the country equipped for her needs. Bridget Spufford had died in May 1989, and her mother, Margaret, had managed to solicit the help of Hawking who had willingly agreed to be a patron of the charity.

The Hawking name carried weight, and an appeal to raise £600,000 was launched in a blaze of local publicity. Hawking went on record as declaring that the attitude of the University towards the handicapped was appalling, stating that they were flouting the law by ignoring an Act of Parliament dating back to 1970, which made it illegal not to provide appropriate access to disabled persons. He spoke of his own situation and how the University had ignored his special needs throughout his undergraduate and postgraduate years, installing a ramp at the DAMTP only under duress and after a long battle when he achieved the status of Reader. The situation was so bad in Cambridge, he revealed, that the National Bureau for Handicapped Students advised people with serious disabilities not to consider Cambridge because of inadequate accommodation.

Hawking also helped to establish a dormitory for handicapped students at Bristol University, which upon completion was named Hawking House. On a filing cabinet in his office at the DAMTP stands an abstract sculpture presented to him for his help in getting the dormitory built.

By 1989, royalties from *A Brief History of Time* had begun to flood in, and with global sales in their millions it was obvious that Hawking no longer needed the financial support of charities to enable him to maintain a very comfortable lifestyle, provide for the education of his children and pay for his round-the-clock nursing. He gratefully acknowledged his enormous debt to the foundations who had saved his life. But, as *A Brief History of Time* gradually became what seemed to be a permanent feature in the best-seller list, unexpected storm-clouds of controversy began to gather over a particular passage in the book.

In Chapter 8, 'The Origin and Fate of the Universe', Hawking refers to the events surrounding the formulation of the cosmological theory of inflation, which we described in

Chapter 11. He picks up the story in 1981, on a visit to Moscow, where the Russian physicist Andrei Linde told him of his latest work on inflation. Linde had written a paper on the subject, but Hawking had pointed out a major flaw in the theory which subsequently took the Russian cosmologist several months to sort out before the rewritten version was ready for submission to a journal.

In the meantime, the day after arriving back from Moscow, Hawking had set off for Philadelphia to collect an award from the Franklin Institute, after which he was invited to deliver a seminar. He recounts the story thus:

> I spent most of the seminar talking about the problems of the inflationary model, just as in Moscow, but at the end I mentioned Linde's idea of slow symmetry-breaking and my corrections to it. In the audience was a young assistant professor from the University of Pennsylvania, Paul Steinhardt. He talked to me afterward about inflation. The following February, he sent me a paper by himself and a student, Andreas Albrecht, in which they proposed something very similar to Linde's idea of slow symmetry-breaking. He later told me he didn't remember me describing Linde's ideas and he had seen Linde's paper only when they had nearly finished their own.[3]

When Steinhardt discovered what Hawking had written about him he was undcrstandably furious. The potential damage to his career was immeasurable. At the time, Steinhardt was a junior professor, while Hawking was Lucasian Professor at Cambridge, and widely acknowledged as one of the most eminent physicists in the world. The whole incident was reminiscent of the conflict, early in the eighteenth century, between the relatively unknown mathematician Gottfried Leibniz and Isaac Newton over who had invented the Calculus. However, the inclusion of this passage in Hawking's best-selling book was not the beginning of the story. The arguments had started back in 1982 after a physics workshop organized by Hawking in Cambridge.

Mike Turner and John Barrow, who had been at the workshop, showed Hawking their draft summary of the meeting, and suggested that some remarks about the Linde and Albrecht–Steinhardt discovery of 'new inflation' could be included. Hawking took exception to the proposed co-credit. Instead of confronting Steinhardt or Albrecht directly, he suggested to Turner and Barrow that they either delete their names or add a reference to a Hawking–Moss paper, crediting it with co-discovery of 'new inflation'.

Hawking's reasons for taking this attitude were, first, that he claimed (incorrectly) that the Steinhardt–Albrecht paper had appeared in print a full six months after Linde's; and, second, that he had discussed Linde's theory at a seminar a few months earlier, a seminar which Steinhardt and Albrecht had been to as well. Angered by Hawking's attitude, Turner and Barrow alerted Steinhardt and Albrecht to the conflict and simultaneously decided, at a risk to themselves, not to follow through with Hawking's request.

Steinhardt wrote to Hawking explaining his position, and sent him notebooks and letters which verified that his work had already been under way before Hawking's talk the previous October. He also stated quite categorically that he had, in any case, no recollection of Hawking mentioning Linde's ideas at the seminar. Most of all, Steinhardt was incensed by the fact that Hawking had gone behind their backs, and that if he had doubts about the validity of their work he should have raised the matter openly. He realized that Hawking was causing this dispute not so much to promote his own interests as to support his friend Linde, but this did not in any way excuse his behaviour.

Hawking wrote back to Steinhardt to say that he had meant nothing by his remarks to Turner and Barrow, and that he fully accepted that the Albrecht–Steinhardt work was independent of Linde's. He even concluded his letter with a friendly wish that they might work together on future projects, making it clear that, as far as he was concerned, the matter was closed.

This was in 1982, before Hawking had begun to write *A Brief History of Time*. It came as quite a surprise, therefore, when in 1988, with Hawking's book in the best-sellers list, Steinhardt was informed of the offending passage. By then, Steinhardt had heard rumours that Hawking had mentioned the controversy in private conversations over the years, and had evidently not let the matter lie as he had implied in his letter to Steinhardt in 1982. However, it was the circumstances in which Steinhardt discovered Hawking's continued pursuit of the matter that really caused offence. Steinhardt had requested some information on obtaining a National Science Foundation grant, and it was the funding officer who pointed out the offending section in Hawking's book. Needless to say, there was no further discussion of National Science Foundation grants on that occasion.

Steinhardt had to defend his reputation. Hawking's behaviour was now having a potentially seriously damaging effect on his career. He decided to substantiate his claims about the Drexel seminar by going through his old notes and obtaining independent verification. Instead, he stumbled upon something much more useful – a videotape of the 1981 seminar. Copying the tape with independent witnesses at every stage, he sent a copy to Hawking in Cambridge and a copy to Bantam in New York, by express mail. Several months passed before Hawking responded to Steinhardt's challenge. This time he wrote to say that the offending text in *A Brief History of Time* would be changed in the next edition, and that the publishers had drafted a press release to announce the change. However, he neither apologized to Steinhardt for the damage his actions had caused nor suggested that his original version had been in any way wrong. It was only after several of Hawking's colleagues around the world began to make it clear they thought he was wrong that he relented.

Chief among Steinhardt's supporters was Mike Turner at Fermilab. He found himself in a very awkward position over the whole affair. He was friendly with both men, but saw Hawking's actions as unjust. Finally, at a meeting in Santa Barbara in 1988, Hawking encountered Turner and asked, 'Are you ever going to speak to me again?' Still angry over the incident, Turner suggested that Hawking could do more to salve the wounds he had caused. In an effort to lay the matter to rest Hawking wrote a letter to *Physics Today*, which was published in the February 1990 issue, in which he said he was sure that the two teams had been working independently on new inflation, and that he was sorry if his account of the incident had been misinterpreted by the readers of his book.

As far as both parties are concerned, the matter is now closed, but Hawking's behaviour on this occasion was patently wrong. The darker aspect of his famous stubbornness had overridden fairness. Steinhardt is still smarting from the incident, which has undoubtedly and quite wrongly damaged his career and caused him totally unnecessary emotional distress. However, as evidenced by the Leibniz–Newton conflict, such disagreements and wrangles are far from uncommon in the history of science. Characters like Hawking do keep the world of science alive and energized by their ideas and imaginations, but the less creative aspects of such strong personalities can sometimes head off at personal tangents with an intensity parallel to their more creative contributions.

Within weeks of *A Brief History of Time* entering the American best-seller list, the film rights for the book were snapped up. An ex-ABC news producer by the name of Gordon Freedman was quick to see the potential of Hawking's book as a film. He also happened to share the same agent as Hawking, Al Zuckerman. Freedman and Zuckerman did a deal and the film rights were sold.

The problem for Freedman was what he was then going to do with the acquisition. He did not want to make a straight documentary of Hawking's life and work – there had been too many of these already, and they had covered the ground quite effectively. On the other hand, he felt there was plenty of scope in the ideas described in the book to produce a film which explored the more esoteric aspects of Hawking's work as well as getting across the essential human-interest angle. A series of coincidences then occurred which eventually led to a viable project.

Freedman went to Anglia Television in Britain. Anglia is based in Norwich, which is close enough to Cambridge for Hawking to be considered a local celebrity. Only a matter of weeks earlier an Anglia TV producer, David Hickman, had approached the commissioning editors with the idea of making a film about Stephen Hawking. Rival broadcasters at BBC East, also based in Norwich, had made the award-winning *Master of the Universe*, and Hickman thought that they should make a programme which tackled the subject in a different way from that of the BBC team. Stirred by the offer from Freedman in the States and by Hickman's proposal, Anglia became interested in the concept and agreed to take on the Freedman project with Hickman as producer and Gordon Freedman as executive producer.

A year passed, during which the producers worked out how they would raise the finances for their project. The original concept was a large-budget TV special, a '*super-Horizon*',* as Hickman described it. For that they would need big bucks. After lunch in London with Caroline Thomson, then commissioning editor of Science Programmes at Channel 4, the network expressed interest in the project, but could not foot the entire bill. At this point Freedman decided to try the big broadcasters in the States. Instead of approaching

* *Horizon* is a science documentary series on British TV.

them directly, he went first to Steven Spielberg's company, Amblin Entertainment, in Los Angeles.

Spielberg had been following Hawking's work for many years and, with an eye on the commercial worth of the project, was immediately interested in the idea of helping to increase public awareness of what Hawking was trying to say in *A Brief History of Time*. Spielberg is another of those who sees Hawking as the late twentieth century's answer to Albert Einstein, and has felt a deep fascination with things extra-terrestrial from a very early age. It was Spielberg's involvement which really brought the scheme into high profile and secured the essential finances needed to bring the project to fruition.

Spielberg and Hawking actually met early in 1990 on the Universal lot at Amblin Studios in Los Angeles, where they posed together for photographers and chatted for over ninety minutes in the Californian sunshine. Expressing a mutual admiration, they apparently got on very well. Hawking had enjoyed *E.T.* and *Close Encounters of the Third Kind*. He even suggested jokingly that their film should be called *Back to the Future 4*. For his part, Spielberg had been greatly taken by *A Brief History of Time*. According to one journalist, observers at the meeting reported that it was Hawking who was the centre of attention – quite a feat in Hollywood, where Spielberg is perceived as a demigod.

In the same month Freedman had contacted Amblin, a film-maker by the name of Errol Morris had approached them with an idea for a new film. Morris had written and directed the critically successful and controversial *The Thin Blue Line*, a film about an alleged cop-killer who was wrongly imprisoned after an incident in Dallas. Morris's idea was to make a film about the mystery surrounding what had happened to Einstein's brain after his death. When the Hawking proposal turned up, Spielberg suggested that Morris might like to look at the idea with a view to directing the project.

Morris had been aware of Hawking's work since his student days, when he had studied philosophy of science at Princeton, and had attended lectures given by the eminent American physicist John Wheeler, who had first applied the term 'black hole' in an astronomical context. David Hickman has suggested that Morris was also interested in the project because, at a certain level, he saw parallels between Randall Adams, the protagonist in *The Thin Blue Line*, and Stephen Hawking. Adams was trapped in a situation which was entirely out of his control, caught up in a web of events over which he had little influence. In the same way Hawking, trapped in a crippled body, is physically ensnared but has mentally transcended this barrier to achieve greatness. Morris is inherently fascinated by such themes, and uses them as a jumping-off point for his iconoclastic movies.

By the end of 1989, with Spielberg's involvement, NBC in America had become interested. The president of the Entertainment Division of the network was a great admirer of *The Thin Blue Line*, and was sold on the idea almost immediately. NBC eventually became the film's major financial contributor. With the interest of two networks under his belt, Freedman then decided to try Japanese television. The idea of a TV special about Hawking backed by Spielberg was very appealing to the Japanese, and Tokyo Broadcasting took very little convincing. The project now had the funding it needed. Between the three networks, the producers had a budget of three million dollars. They could effectively make the film they wanted.

Errol Morris's approach was to build the film around a series of interviews, recording much more footage than is used in the final version. Cutting this interview material to perhaps half its original length, he then began to construct visual images around what remained. In the first stage of the project, researchers drew up a list of Hawking's friends, family and colleagues from around the world who they thought might be interested in taking part in the project. However, they were soon surprised to discover that there were many people who did not want to be in the film.

Hickman believes there is some resistance to media people in Cambridge. Like Peter Guzzardi, he felt that some of Hawking's students – as well as more senior colleagues – resented the idea of serious scientific work being over-simplified. He also detected that, despite the runaway success of *A Brief History of Time*, there was a definite closing of ranks in certain quarters at the suggestion of a commercial film being made around Hawking's ideas.

'Cambridge University is a very tight community,' he said. 'There are numerous rivalries, jealousies, animosities. Despite the fact that the interviews were totally unscripted (they could talk about what they had for breakfast if they wanted), there was an undoubted feeling that we were a *News of the World* on screen.'

Fortunately for the producers, however, there were plenty more interested participants than those suffering from delusions that they were being coerced into something slightly unsavoury.

In January 1990, sound stages at Elstree Studios were block-booked for two weeks. The first people to move in were the set-designers. Morris had the idea that he would give the designer the name of an interviewee and a rough idea of his or her relationship with Hawking, and the designer would then go away and create individual sets for each interviewee to be filmed in. Sometimes the set had absolutely no relevance to the subject; for other interviews it matched the topic of the interview.

As the interviews were unscripted, Morris would often say to the interviewee, 'Look, I don't really know how to start this interview. Why don't you just tell me some stories?' He has what he calls the two-minute rule: 'If you give people two minutes, they'll show you how crazy they are.'

For *A Brief History of Time* they conducted over thirty interviews in thirteen days at Elstree, using thirty-three different sets. Interviewees included Dennis Sciama, Dr Robert Berman, Isobel Hawking, friends from school and undergraduate days and co-workers at the DAMTP such as Gary Gibbons. However, star billing was reserved for Stephen Hawking himself.

The most important set at Elstree during the fortnight of filming was a reconstruction of Hawking's office at the DAMTP. No effort was spared in recreating the room in intimate detail. Even Hawking was bemused by Morris's attention to minutiae.

'I'm surprised they went to all that trouble because most people wouldn't have known if it had been different,'[4] he said.

Morris had wondered about Hawking's fascination with Marilyn Monroe. Hawking smiled, and explained that he had very much enjoyed *Some Like It Hot*, and ever since his family and friends had insisted on buying him Marilyn merchandise at every opportunity: posters from Lucy and his secretary, a Marilyn bag from Timothy, and a towel from Jane. 'I suppose you could say she was a model of the Universe,'[5] he had joked.

Morris had also decided to have built a reproduction of Hawking's wheelchair, accurate to the last detail of the licence plate, for when he could not make a shoot. Using 'macro-filming' techniques, he could get extreme close-ups of the chrome-work and leather, filling the screen as an image to accompany an interview on voice-over. According to Hickman, Hawking's childhood home at 14 Hillside Road was filmed almost brick by brick.

Hawking himself was shot against a blue-screen so that his image could be projected on to any backdrop the director chose. The original intention was to have Hawking narrate relevant parts of the film using his voice synthesizer. However, it soon became clear that the harshness of the voice was irritating after a while when used as a voice-over. Consequently, Morris decided against the idea and the viewer hears Hawking's voice only when he is actually talking to camera. The use of blue-screen filming gives the director enormous flexibility. 'I can place Stephen Hawking where he belongs, in a mental landscape rather than a real one,'[6] Morris has said.

What the viewer does not see, however, is astronauts falling into black holes or other such science-documentary clichés. As Hickman points out, 'No one has seen a black hole – they are theoretical objects as far as we know. The subject-matter of this film lies in the realms of the imagination.'

With a three-million-dollar budget, Hickman, Freedman and Morris could call on the very best people in the business to handle design, lighting, cinematography, sound production and other essential technical support. The background staff responsible for transforming Morris's ideas into a viable product had impeccable credentials; between them they had worked on over a dozen major Hollywood films, including *Edward Scissorhands*, *Batman II*, *American Gigolo* and *Wild at Heart*. The American composer Philip Glass was commissioned to write the film score, his polyrhythmic electronic music acting as a perfect complement to Morris's visual acrobatics.

Hickman says that the film is really about God and time and not so much about scientific investigation or Hawking's disabilities:

We are far more interested in the concepts Stephen has tried to portray in his book than in producing a straight science documentary asking questions like 'What is the future of cosmology?' The most exciting thing about cosmology is the fact that it interfaces metaphysics and conventional science. It's very interesting that Stephen has attracted a lot of attention over the religious aspects of his work, as well as the fact that he is close to a number of physicists with deep theological concerns, such as Don Page.

On the days when Hawking was called upon for shooting he travelled to Elstree with his team of nurses and aides in the specially converted VW van he acquired soon after receiving the cash award that came with the Wolf Prize. On the set, a reverent hush regularly descended on the crew and technicians. Hawking, despite his disabilities, commands a powerful presence which surprises most people on their first meeting. Seated in his wheelchair he would spend hours under the studio lights, silently observing the frenzy of activity around him as the camera zoomed in for a close-up, or make-up people dabbed rouge on his cheeks between takes.

The filming of *A Brief History of Time* was completed in spring 1990, but Morris's film-making technique is labour-intensive during the editing stage of a project. This took up the rest of 1990 and the early part of 1991, and the film was finally to hit cinemas in America and Europe in the spring of 1992. The intention was to show the movie in selected theatres for a short period and then for it to be networked internationally by the broadcasters who financed the project, NBC in the States, Tokyo Broadcasting in Japan and Channel 4 in the UK. It was then sold to other broadcasters around the world and destined ultimately to appear in the stores as a video.

While the movie project was in the editing stage, during the summer of 1990, the seemingly unthinkable happened. Shock-horror headlines appeared in a number of national newspapers announcing the sad fact that Stephen and Jane Hawking had separated after twenty-five years of marriage.

In fact, the two of them had been growing apart for a number of years. As Hawking's career reached new heights of fame and success, the awards and medals piling up along with honours from all parts of the world, Jane had felt increasingly isolated. She had begun to accompany Stephen on foreign trips far less frequently, and as she no longer had the responsibility of nursing her husband, she had started to turn her attention towards her own interests – her work, her garden, her books and an increasingly active involvement in one of the best choirs in Cambridge.

The academic community in Cambridge was shocked by the news. For as long as anyone could remember, Stephen had taken great pains to promote the role Jane had played in his life and, despite their disagreements, to outsiders their marriage was a model of security. For weeks friends and colleagues were plagued by newspaper reporters who had staked out the Hawkings' home in West Road in an attempt to get a scoop and dig the dirt on the marriage break-up. Hawking was a world-famous figure, and in the minds of the Sunday rag editors there was the macabre twist of Stephen's disability to mix into a centre-page splash.

Thankfully, the gutter press never succeeded in finding the angle they wanted. In Cambridge, the scientific community closed ranks and family friends, if they knew any details about why the couple had parted, were saying nothing. Gradually, however, stories began to emerge. There were rumours of extramarital relations developing over a number of years long before their marriage had reached crisis-point; but those who knew the couple well regarded as far more significant tales of increased tensions between Stephen and Jane over the old religious arguments. Their disagreements had been swept under the carpet for many years, but with the writing of *A Brief History of Time*, it appears that the wounds had been re-opened.

Through his work, Hawking's early agnosticism had become more overtly atheistic, and with his no-boundary theory he had effectively dispensed with the notion of God altogether. Yet, ironically, Jane's deeply held religious convictions had been one of the strengths which had enabled her to cope so well with the burden imposed by Stephen's increasing disability. However, the couple had lived with religious disunity for most of their married life, so that on its own was certainly insufficient reason to separate.

As reported in a number of newspaper articles, the break came when Stephen left Jane to move into a flat to live with the nurse who had looked after him for a number of years, Elaine Mason. According to reports, as Stephen and Jane had drifted apart, he and Elaine had grown closer. For a number of years it was Elaine rather than Jane who had accompanied him on his foreign travels and with whom he spent much of his working life. The situation was complicated by the fact that Elaine was married to David Mason, the computer engineer who had adapted Hawking's computer so that it could be fitted to his wheelchair. The couple had two children, and in fact David Mason and Hawking had met at the gates of the primary school which both Timothy and the Mason boys had attended. It was through this initial contact, and Hawking's request for a chair-mounted computer, that Mason had been able to start his own computer business and Elaine Mason had later become one of Stephen's team of nurses.

Jane had cared for her husband for over twenty-five years, sacrificing many of her own personal hopes and ambitions along the way, but as fame and international success had begun to take over his life, and their paths diverged, it appeared that they no longer needed each other. Some commentators have tried to place the blame on Stephen, but many others believe that such views are wide of the mark. In any marriage break-up, blame is not a word to use lightly. Certainly, Jane has devoted most of her life to Stephen, almost single-handedly taking care of him when he was a little-known physicist struggling to overcome disability and develop his career. However, things change; many married couples grow apart from each other. A number of friends feel that Stephen should not be blamed for leaving the woman who had done so much for him. It is an insult to Jane's dedication and commitment for others to place the past like a yoke around his neck.

Like all break-ups, theirs caused a great deal of sadness. The Hawking children took the news particularly badly. Robert, then twenty-three, had graduated in physics from Cambridge the previous year and was already embarking on postgraduate work; Lucy, nearly twenty, was at Oxford University studying modern languages. The two of them, though naturally upset, were old enough to accept the situation and were developing their own lives away from home, carving out their own independence. The separation hit the youngest, Timothy, the hardest. Then barely eleven, he was too young to understand fully the reasons why his father had left their home in West Road.

There is little doubt that the trauma of separation had affected Stephen as much as any of those involved, and reporters claimed that the famous Hawking smile was now rarely seen. Others pointed out that, at the time, he was displaying great emotional swings. He could be outwardly very happy for a while, smiling and joking with his colleagues and students, and then fall into a depression, casting a mournful shadow over the atmosphere at the DAMTP.

It is important to remember that although a great many people go through similar emotional unheavals, the vast majority of them have a number of advantages over Stephen Hawking. There are ways in which their emotions can be diverted and released; ineffectual as these methods often prove to be, they were not available to him at all. He could not scream and shout, go for a run or indulge in a drinking binge, he could not smoke himself stupid or even speak to friends with ease. And, although it was he who made the break, the pain was undoubtedly still there.

Many people who claim to know Stephen Hawking have

been over-protective towards him, especially since the announcement of the separation. This attitude is misguided, and is usually shown by people who turn out not to know him at all well. Close friends know that Hawking needs nobody to protect him – he is perfectly capable of looking after himself. The same people who try to protect Stephen also make the mistake of attempting to imbue him with feelings and emotions different from those of the rest of us, almost as if, because of his highly tuned intellect, he did not share the same dreams, hopes and passions that the rest of humanity experiences.

One of his closest friends, David Schramm, has known Hawking for over twenty years and has little patience with those who try to create an image of Stephen as in any way emotionally different from others. He has never pulled any punches when it comes to his friend's personal life. He once introduced Hawking at a talk he gave in Chicago, by saying, '. . . as evidenced by the fact that his youngest son Timothy is less than half the age of the disease, clearly not all of Stephen is paralysed!' Apparently half the audience were shocked speechless, but Hawking loved it.

Schramm believes that people are scared to face the fact that, in emotional terms, Stephen Hawking is a normal man. Because of the power of his intellect as well as the singular nature of his physical condition, they convince themselves that he does not feel the same way as others. Stephen loves the company of women, he enjoys flirting, he appreciates physical beauty: why else would he have a poster of Marilyn Monroe in his office? Probably not for her intellect. Hawking's relationship with Elaine Mason is not one based on pity or other such feeble foundations. According to Schramm, who has spent a lot of time with the couple, there is a genuine love between them.

Hawking refuses to talk publicly about his private life, and makes that a stipulation of any interview these days. The journalists, for their part, continue to speculate on the causes and outcomes of the split. Jane, for her own reasons, is equally tight-lipped on the matter. She refuses to talk to anyone about anything to do with her and Stephen. She turned down repeated requests from the producers to take part in the film of *A Brief History of Time*, and she agrees to participate in interviews only with journalists she knows personally. She will not be drawn on anything even vaguely relating to her life with Stephen.

Pictures of Jane and the children still decorate Hawking's office at the DAMTP, but the separation is without doubt an acrimonious one. Friends claim that Jane speaks bitterly about it. She is now under no obligation, as one acquaintance put it, to 'promote the greater glory of Stephen Hawking'.[7] Only a year earlier, Jane had told a reporter that 1989 had been the year when everything had fallen into place for them, when they had reached a new high point in their lives:

> For me the fulfilment stems very much from the fact that we have been able to keep going, that we have been able to remain a united family. The awards were like the sugar frosting on the cake. I wouldn't say that is what makes all the blackness worth while. I don't think I am ever going to reconcile in my mind the swings of the pendulum that we have experienced in this house – really from the depth of a black hole to all the glittering prizes.[8]

She explained to another journalist that her role was no longer to look after a sick man but 'simply to tell him that he's not God'.[9] Perhaps in such statements as this the murmurings of deep-rooted resentments and disquiet can be detected. Yet in the concluding scene of the BBC's *Master of the Universe* programme we see Stephen and Jane looking down on a sleeping Timothy in their house in West Road while Hawking's computer voice declares, 'I have a beautiful family, I am successful in my work, and I have written a best-seller. One really can't ask for more.'[10]

The couple still meet occasionally, and Stephen sees Timothy as much as he can. It has been said that, of all his children, Timothy is most like him. They still manage to play games together, Stephen winning at chess while Timmy beats his father at Monopoly. The older children have always known that their father can be a difficult man to live with at times. In the late eighties, Lucy, in the *Master of the Universe* documentary, said:

I'm not as stubborn as him. I don't think I would want to be that stubborn. I don't think I have quite his strength of mind, which means he will do what he wants to do at any cost to anybody else.[11]

By the summer of 1991, *A Brief History of Time* had been in the British best-seller list for 150 weeks, and had notched up millions of sales in twenty different languages. In America, the audiotape version of the book had become the best-selling book tape in the country, and an unabridged version on four tapes had been released in Britain. Hawking was receiving piles of mail each day at the DAMTP, ranging from requests from amateur physicists in India asking for enlightenment about certain passages in the book to letters from entrepreneurs in Middle Eastern states inquiring about publishing rights in their countries.

The combination of publishing demands, the filming of *A Brief History of Time* and the circumstances surrounding his private life led a number of observers to wonder if Hawking's science would suffer, if he had lost touch with academia. Nothing could have been further from the truth. The sign on his office door which reads 'QUIET PLEASE, THE BOSS IS ASLEEP' must be read in the spirit in which it was intended – as a joke. It is safe to say that it was science that kept Stephen Hawking sane during this difficult time. He remains the most likely man to lead the way to formulating a Grand Unified Theory, a 'theory of everything':

I am still trying to understand how the Universe works, why it is the way it is and why it exists at all. I think there is a reasonable chance that we may succeed in the first two aims, but I am not so optimistic about finding why the Universe exists.[12]

There were indeed great pressures on him. He was aware of the fact that, for the time being, he could not commit so much time and effort as he would have liked to pure research and the care of students at the DAMTP. However, somehow he succeeded in juggling his responsibilities and keeping everyone happy. His devotion to science had pulled him through many difficulties in the past. Above all else, Hawking is totally dedicated to physics. It was his first love, and there is little doubt that it will be his last.

Travels abroad during the 1990s include a large number of business trips in connection with *A Brief History of Time* and its spin-offs. There are endless requests to give public lectures to parties of schoolchildren, city dignitaries and interested members of the general public. Hawking always finds it difficult to turn down such invitations.

The symptoms of his condition appeared to slow again after the incident at CERN when he almost lost the battle. The disease progresses in sudden leaps and bounds, then periods of unpredictable duration on a plateau. It is widely felt that one more downward change and it will be the end of the road. However, people have been writing Hawking off for over twenty-five years, and so far he has proved them wrong. He certainly doesn't think about it too much. Ironically, he has never been the type of person to dwell on his own longevity; he takes each day as it comes and makes the most of it.

So, who is Stephen Hawking, the man? He is a force to be reckoned with, of that there is little doubt. His strength of personality is formidable – given his physical condition, how else could he have survived and achieved greatness in more than one arena? He can be ruthless; he drives a hard bargain with life and approaches it head on. He finds it hard to compromise; his force of will can sometimes work against him. Many people find him abrasive, but on the other hand he is famous for his sense of humour. He has many close friends and admirers, and has proved himself to be a loving and affectionate father. It is impossible to know the man's inner thoughts, so intimately linked as he is to machines, a set of cold devices enabling him to move, speak and breathe. His face is, if anything, more expressive than most because, aside from his gift for succinct language, it is just about our only window into his mind.

His work is a major part of Stephen Hawking, but so few of us can understand it except in the vaguest pictorial terms. His attempt to communicate his understanding to the world at large through his best-selling book has succeeded. Of course, a great many copies of *A Brief History of Time* have hardly been opened, left to adorn book-shelves as fashion accessories, but despite this there are many – perhaps millions – who have learnt more about the Universe we all live in through reading his words. With this alone he has achieved astounding success by awakening a sceptical public and even more sceptical media to the beauty of science, a subject at the heart of our society and the future of civilization. The popularization of science has seen a new renaissance, thanks in large measure to his efforts.

Beyond all this, running deeper than his hugely successful venture into literature, beyond even his scientific achievements, there remains the human triumph of his very survival, the strength of his spirit in accomplishing more than most of us dream about. Some claim that Stephen Hawking has made it only because of the unfortunate circumstances he found himself in, but such glibness denies the very essence of humanity. Others crumble under far less strain. It is the Stephen Hawkings of this world who soar, no matter what. To those intent on destroying legend and denigrating achievement, he has a typically modest, but perfectly accurate response. It would stand equally well as his own epitaph and as a philosophy of life for all of us to follow:

One has to be grown up enough to realize that life is not fair. You just have to do the best you can in the situation you are in.[13]

References

Quotations without sources are from interviews with the author.

1 The Day Galileo Died

1. S. W. Hawking, *A Short History* (privately produced pamphlet).
2. Michael Church, 'Games with the cosmos', *Independent*, 6 June 1988.
3. S. W. Hawking, *A Short History*.
4. Michael Church, 'Games with the cosmos'.
5. *Albanian*, May 1958.

3 Going Up

1. S. W. Hawking, *A Short History*.
2. S. W. Hawking, *A Short History*.

4 Doctors and Doctorates

1. S. W. Hawking, *A Short History*.
2. Tony Osman, 'A master of the Universe', *Sunday Times Magazine*, 19 June 1988.
3. S. W. Hawking, *My Experience with ALS* (privately produced pamphlet).
4. S. W. Hawking, *My Experience with ALS*.
5. S. W. Hawking, *My Experience with ALS*.
6. S. W. Hawking, *My Experience with ALS*.
7. S. W. Hawking, *My Experience with ALS*.
8. S. W. Hawking, *My Experience with ALS*.
9. Bryan Appleyard, 'Master of the Universe: Will Stephen Hawking live to find the secret?', *Express News*, San Antonio, Texas, 3 July 1988.
10. Dennis Overbye, 'The wizard of space and time', *Omni*, February 1979, pp. 45–107.
11. S. W. Hawking, *A Short History*.

6 Marriage and Fellowship

1. S. W. Hawking, *A Short History*.
2. S. W. Hawking, *A Short History*.
3. John Boslough, *Beyond the Black Hole: Stephen Hawking's Universe*, London, Fontana, 1985.
4. Bryan Appleyard, 'Master of the Universe'.
5. 'Bob Sipehen, The sky's no limit in the career of Stephen Hawking', *West Australian*, 16 June 1990.
6. *20/20*, ABC Television Broadcast, 1989.
7. Ellen Walton, 'A brief history of hard times', *Guardian*, 9 August 1989.
8. Dennis Overbye, 'The wizard of space and time'.
9. Michael Harwood, 'The Universe and Dr Hawking', *New York Times Magazine*, 23 January 1983.
10. Dennis Overbye, *Lonely Hearts of the Cosmos*, London, Macmillan, 1991.

8 The Breakthrough Years

1. Jerry Adler, Gerald C. Lubenow and Maggie Malone, 'Reading God's mind', *Newsweek*, 13 June 1988.
2. Stephen Hawking, *A Brief History of Time*, London, Bantam, 1988.
3. Dennis Overbye, *Lonely Hearts of the Cosmos*.
4. Dennis Overbye, *Lonely Hearts of the Cosmos*.
5. Ian Ridpath, 'Black hole explorer', *New Scientist*, 4 May 1978, p. 307.
6. John Boslough, *Beyond the Black Hole*, p. 25.
7. Timothy Ferris, 'Mind over matter', *Vanity Fair*, June 1984.
8. Dennis Overbye, *Lonely Hearts of the Cosmos*.

9 When Black Holes Explode

1. S. W. Hawking, B. Carter and J. Bardeen, *Communications in Mathematical Physics*, 1973, Vol. 31, pp. 161–70.
2. Stephen Hawking, *A Brief History of Time*, p. 105.
3. S. W. Hawking, *Scientific American*, January 1977, pp. 34–40.
4. S. W. Hawking, *Nature*, 1974, Vol. 248, pp. 30–31.
5. J. Taylor and P. Davies, *Nature*, 1974, Vol. 250, pp. 37–8.

10 The Foothills of Fame

1. S. W. Hawking, *My Experience with ALS*.
2. Dennis Overbye, *Lonely Hearts of the Cosmos*.
3. Dennis Overbye, *Lonely Hearts of the Cosmos*.

4. Alan Lightman and Roberta Brawer, *Origins: The Lives and Worlds of Modern Cosmologists*, Harvard University Press, 1990, p. 406.
5. Michael Harwood, 'The Universe and Dr Hawking'.
6. Dennis Overbye, *Lonely Hearts of the Cosmos.*
7. Bryan Appleyard, 'Master of the Universe'.
8. Timothy Ferris, 'Mind over matter'.
9. John Boslough, *Beyond the Black Hole*, p. 25.
10. Timothy Ferris, 'Mind over matter'.
11. Ellen Walton, 'A brief history of hard times'.
12. Ellen Walton, 'A brief history of hard times'.
13. *Master of the Universe*, BBC TV Broadcast.
14. Ellen Walton, 'A brief history of hard times'.
15. *Master of the Universe*, BBC TV Broadcast.
16. *Master of the Universe*, BBC TV Broadcast.
17. *Master of the Universe*, BBC TV Broadcast.
18. *Master of the Universe*, BBC TV Broadcast.
19. *20/20*, ABC Television Broadcast.
20. Michael Harwood, 'The Universe and Dr Hawking'.
21. Stephen Hawking, *A Brief History of Time.*
22. Bryan Appleyard, 'Master of the Universe'.
23. Jeremy Hornsby and Ian Ridpath, 'Mind over matter', *Sunday Telegraph Magazine*, 28 October 1979.
24. Kitty Ferguson, *Stephen Hawking: A Quest for the Theory of Everything*, New York, Franklin Watts, 1991.
25. Stephen Hawking, *A Brief History of Time.*
26. D. Page, 'Hawking's timely story', *Nature*, 1988, Vol. 333, pp. 742–3.
27. Stephen Hawking, *A Brief History of Time.*
28. Stephen Hawking, *A Brief History of Time.*
29. Bryan Appleyard, 'Master of the Universe'.
30. Dennis Overbye, *Lonely Hearts of the Cosmos.*
31. Dennis Overbye, *Lonely Hearts of the Cosmos.*

11 Back to the Beginning

1. Stephen Hawking, *A Brief History of Time*, pp. 140–41.

12 Science Superstardom

1. *Cambridge Evening News*, 31 January 1978.
2. John Boslough, *Beyond the Black Hole*, p. 28.
3. Michael Harwood, 'The Universe and Dr Hawking'.
4. Michael Harwood, 'The Universe and Dr Hawking'.

5. Dennis Overbye, 'The wizard of space and time'.
6. Stephen Shames, 'Stephen Hawking: A thinking kind of hero', 1988.
7. *Sunday Telegraph Magazine.*
8. Tony Osman, 'A master of the Universe'.
9. Colin Wills, 'Triumph of mind over matter', *Sunday Mirror*, 4 September 1988.
10. 'The sky's no limit in the career of Stephen Hawking', *West Australian*, 1989.
11. Timothy Ferris, 'Mind over matter'.
12. Dennis Overbye, *Lonely Hearts of the Cosmos.*
13. Dennis Overbye, *Lonely Hearts of the Cosmos.*
14. Stephen Shames, 'Stephen Hawking: A thinking kind of hero'.
15. John Gribbin, *In Search of the Big Bang*, London, Heinemann, 1986, pp. 387–8.

13 When the Universe has Babies

1. E. Fahri and A. Guth, *Physics Letters*, 1987, Vol. 183 B, pp. 149–53.
2. Stephen Hawking, *A Brief History of Time*, p. 137.

14 *A Brief History of Time*

1. 'Book news', *Bookseller*, 21 October 1988.
2. 'Book news', *Bookseller*, 21 October 1988.
3. John Boslough, *Beyond the Black Hole*, p. 26.
4. John Boslough, *Beyond the Black Hole*, p. 27.
5. Leonore Fleischer, 'Talk of the trade', *Publishers Weekly*, 15 January 1985.
6. Ellen Walton, 'A brief history of hard times'.
7. 'Top city scientist taken to hospital', *Cambridge Evening News*, 17 August 1985.
8. Ellen Walton, 'A brief history of hard times'.
9. Ellen Walton, 'A brief history of hard times'.
10. Kitty Ferguson, *Stephen Hawking: A Quest for the Theory of Everything.*
11. 'Book news', *Bookseller*, 21 October 1988.
12. 'Book news', *Bookseller*, 21 October 1988.
13. 'Book news', *Bookseller*, 21 October 1988.
14. 'Book news', *Bookseller*, 21 October 1988.
15. Charles Oulton, 'Cosmic writer shames book world', *Sunday Times*, August 1988.
16. Charles Oulton, 'Cosmic writer shames book world'.
17. 'Book news', *Bookseller*, 21 October 1988.

18. Denise Housby, *Cambridge Evening News*, 30 August 1988.
19. John Maddox, 'The big bang book', *Nature*, 1988, Vol. 335, p. 267.
20. Simon Jenkins, 'A dance to the music of imaginary time', *Sunday Times*, 28 August 1988.
21. John Maddox, 'The big bang book'.
22. 'Up and down the city road', *Independent Magazine*, 27 April 1991.
23. Letters page, *Independent Magazine*, 4 May 1991.
24. Letters page, *Independent Magazine*, 4 May 1991.
25. Simon Jenkins, 'A dance to the music of imaginary time'.

15 The End of Physics?

1. Stephen Hawking, *Newsweek*, 13 June 1988.
2. M. Green, *Scientific American*, September 1986, pp. 44–9.

16 Hollywood, Fame and Fortune

1. Tim Verney, 'Top cash prize for brilliant city academic', *Cambridge Evening News*, 21 January 1988.
2. Alan Kersey, 'Musical tribute to brave professor', *Cambridge Evening News*, June 1989.
3. Stephen Hawking, *A Brief History of Time*, 1st edition, New York, Bantam, 1988.
4. David Gritten, 'A brief movie of time', *Sunday Correspondent*, 1990.
5. David Gritten, 'A brief movie of time'.
6. James Delingpole, 'Limelight', *Evening Standard*, 27 June 1990.
7. Nigel Hawkes, 'Defying the gravity of physics', *The Times*, 27 October 1990.
8. Pauline Hunt, 'Glittering triumph of an inspiring family', *Cambridge Evening News*, 19 July 1988.
9. Tony Osman, 'A master of the Universe'.
10. *Master of the Universe*, BBC TV Broadcast.
11. *Master of the Universe*, BBC TV Broadcast.
12. *Master of the Universe*, BBC TV Broadcast.
13. *20/20*, ABC Television Broadcast.

Index